TRAIN WRECK

TRAIN ✴ WRECK

The
FORENSICS *of*
RAIL DISASTERS

GEORGE BIBEL

The Johns Hopkins University Press
Baltimore

© 2012 Johns Hopkins University Press
All rights reserved. Published 2012
Printed in the United States of America on acid-free paper

Johns Hopkins Paperback edition, 2018

9 8 7 6 5 4 3 2 1

Johns Hopkins University Press
2715 North Charles Street
Baltimore, Maryland 21218-4363
www.press.jhu.edu

The Library of Congress has cataloged the hardcover edition of this book as follows:
Bibel, G. D. (George D.)
    Train wreck : the forensics of rail disasters / George Bibel.
        p. cm.
    Includes bibliographical references and index.
    ISBN 978-1-4214-0590-2 (alk. paper) — ISBN 978-1-4214-0652-7 (electronic)
— ISBN 1-4214-0590-3 (alk. paper) — ISBN 1-4214-0652-7 (electronic)
    1. Railroad accidents—Investigation. I. Title.
    HE1779.B535 2012
    363.12'265—dc23   2011048389

A catalog record for this book is available from the British Library.

ISBN-13: 978-1-4214-2707-2
ISBN-10: 1-4214-2707-9

*Special discounts are available for bulk purchases of this book. For more information,
please contact Special Sales at 410-516-6936 or specialsales@press.jhu.edu.*

Johns Hopkins University Press uses environmentally friendly book materials,
including recycled text paper that is composed of at least 30 percent post-consumer
waste, whenever possible.

To my lovely bride, Mary Pat,
who waited patiently for this train's caboose

CONTENTS

ACKNOWLEDGMENTS

I would like to acknowledge help from Mark Aldrich. His book, *Death Rode the Rails: American Railroad Accidents and Safety, 1828–1965,* was an important historical reference, as were his many emails. I would also like to acknowledge Larry Schlosser for patiently explaining railroad operations and Richeldis Nelson and Ben Waldera for extensive editing.

TRAIN WRECK

1

The Railroad Industry
(as Seen through Accidents)

This book addresses three simple questions: why trains crash, what happens when they do, and what safety improvements have been made in response to accidents. These questions are addressed by discussing numerous examples of train wrecks that are further explained using high school science. Selected accidents have an interesting story and a well-documented explanation of the underlying science. This is a train wreck book that teaches science (or a science book that tells train wreck stories).

Although not a book about high-speed trains, the accidents and safety principles developed are fundamental to the safe operations of all trains. The current safety rules (often based on tragedy) exist for a reason and are better understood with historical and scientific explanations.

BEFORE LAUNCHING INTO AN ENTIRE BOOK about how trains crash, I should start with an apology to the entire industry and first explain how trains mostly do not crash. An average freight train replaces 280 trucks and the potential human error associated with 280 drivers.[1] Trains with self-steering wheels are fundamentally safer than anything with rubber tires, and statistics bear this out. Extensive safety systems continue to evolve, and, unlike most car drivers, accident-prone railroaders are fired.

Even when the statistics were grim, the alternatives were far worse. Just as riding a train is safer than driving a car today, a train ride was always safer than the competition. In 1867 (with far fewer people and vehicles), about four pedestrians were killed per week in New York City, somewhat fewer than today. By one estimate, stage coaches were 60 times more dangerous than riding a train in the nineteenth century.

Today a fatality occurs in autos about every 100 million passenger miles. Over a recent 5-year period, passengers in trains were 19 times safer. In fact, passenger train fatalities are so rare that one bad accident alters the statistics. Eliminating the single worst U.S. passenger train accident in the twenty-first century (see Chapter 2), passengers in trains are 45 times safer than those in cars.

Up until the last generation or so, trains have been safer than planes. The statistics could be reinterpreted; 94% of all commercial plane accidents occur during take-off, climb, approach, and landing. Flight accidents per mile of those phases of flight would suggest different conclusions about the safety of trains versus planes.

MAJOR HISTORICAL TRENDS

The history of railroading and of railroad safety has been dominated by major economic trends. The first federal regulatory agency, the Interstate Commerce Commission (ICC), was created in 1887 to police railroad shipping rates. At that time the railroads were rich, powerful, and potentially monopolistic. Since their peak in 1920, the railroads have been in massive decline. In 1920, passenger trains carried more than 47 billion passenger miles and 75% of all intercity freight. After the boom of World War II, the 1950s through the 1970s was a particularly difficult period for the railroad industry. By 1980, the population had nearly doubled, yet passenger traffic had declined by 85%. And the railroads' share of intercity freight traffic had declined from 75% to 37% (1920 versus 1980).

It wasn't just the obvious competition from cars, trucks, and planes. The use of coal, which once heated almost every building, including homes, and powered 60,000 locomotives, is down by almost two-thirds. Today gasoline and crude oil are mostly transported through pipelines. Traffic on inland waterways increased by 150% between 1950 and 1980, and decreased manufacturing also reduces freight traffic.[2] In one of the ironies of railroading, the railroads pay real estate taxes on their track to subsidize highways, airports, and river-dredging.

According to the U.S. General Accounting Office, poor labor productivity also played a role in the decline of the railroads. For example, the positions of fireman (to tend the boiler in a steam locomotive) and telegraph operator existed long after both had become obsolete. Long after steam locomotives disappeared in the mid-1950s, legal wrangling about the fireman's position made it to the U.S. Supreme Court in 1963 and 1968.

Quoting the U.S. Department of Transportation in 1978, "The current system of railroad regulations reflects a series of uncoordinated actions . . . The result is a hodgepodge of inconsistent and often anachronistic regulations that no longer correspond to the economic conditions of the railroads, the nature of intermodal competition, or the often conflicting needs of shippers, consumers and taxpayers."[3] Population shifts to the west and south made tracks in the Midwest and along the East Coast obsolete. Track abandonment and/or rate change proceedings could take years; meanwhile, that route was forced to operate at a loss.

In 1976, 11 of 36 Class I railroads were earning a negative rate of return and 3 were bankrupt.[4] The overall return on investment for all railroad companies was about 2%. With billions of dollars of deferred maintenance, the railroads could not afford to keep their track and equipment in good repair. In the late 1970s, the Federal Railroad Administration (FRA) created a new accident category, the "standing derailment," for cars that derailed while standing still.

In 1971, Amtrak was formed and operated by the federal government to take over bankrupt intercity rail traffic. Today Amtrak remains highly subsidized except for the successful fast trains on the densely populated Northeast Corridor between Washington, D.C., and Boston. The remaining regional and commuter passenger trains have become subsidized entities owned by local governments.[5]

Freight railroads wanted the unregulated freedom to upgrade routes to compete, negotiate contracts in secret, and above all set prices to what the market would bear. Intermodal freight traffic was an innovation of the 1970s. To reduce handling, damage, and costs, the same container was used on ships, barges, trains, and trucks. Just as ships needed to reconfigure for container transport (entire fleets became obsolete), so did the railroads. Updating a train route for increased speed and tonnage involves upgrading hundreds of miles of track, roadbed, bridges, and signals. The railroads could not justify the expenditure until they negotiated long-term contracts—still banned per government regulations in 1980.

Railroads have always been (and will always remain) the cheapest form of land transportation. In fact, today they like to brag about shipping 1 ton of freight 480 miles (772 km) on 1 gallon of fuel, about 4 times better than trucks. The problem for railroads has always been delivery schedules. One major railroad has more than 13,000 origin-destination pairs. The average number of cars for each pair is only 11 per month. Freight cars accumulate in classification yards. When enough cars are collected, they are attached to a train going in the right direction. The process starts over again at the next yard. Without any special priority, freight cars average about 50 miles (80.5 km) per day. The exact opposite scenario is a dedicated unit train loaded with one product. This train proceeds at 50 mph (80.5 km/h), bypasses all yards, and stops only to change crews and take on fuel.

A custom contract can optimize pricing, financing, delivery schedule, guaranteed volume, and damage limits for each customer. Damage is a difficult business/technical problem that creates complex contractual obligations. Consider placing 20 dozen eggs in a wheelbarrow and racing down a bumpy street. Now consider negotiating a contract guaranteeing the number of unbroken eggs delivered. With long-term contracts the railroad could afford to design and build custom freight cars (for each product) to minimize damage.

THE STAGGERS ACT OF 1980

In 1980, Congress passed the Staggers Act, allowing railroads to decide which routes to upgrade (and which to shut down or sell off), set their own rates based on market demand, and negotiate secret contracts. In a few years there were more than 30,000 negotiated contracts with discounts for higher volume or for long-term guaranteed volume as well as short-term discounts during economic slowdowns. With improved service the railroads could better compete with trucks. And railroad companies like to brag about improved safety and service since deregulation.

The freight railroads have invested $500 billion in track and equipment since 1980. (Among trucks, barges, airplanes, passenger trains, and freight trains, only freight trains spend billions to install and maintain their infrastructure.) Increased operating efficiencies have actually decreased inflation-adjusted rates (from 1980 to 2010) by almost 50%. Since 1980, accidents and injuries have declined by 72% and 82%, total ton-miles shipped have doubled, and tons shipped per employee have increased by a factor of 5. Today the American freight train system, by any measure, is universally recognized as the safest and most efficient in the world.

ACCIDENTS TODAY
Collisions

Collisions are usually caused by human operators missing a signal to stop. Only 5% of all accidents are collisions, but they account for 80% of fatalities. Collisions are simply more dangerous.[6] Today most train collisions in the United States involve freight trains, because most trains carry freight.

Over the 10-year period 1999–2008, train collisions in the United States averaged 216 per year. Most collisions are minor, with 77% occurring in yards. All collisions risk worker injury and possible release of hazardous wastes. Collisions in yards and sidings, however, are at lower speeds, have less potential for damage and injury, and interact less with civilians. High-speed collisions are usually followed by a subsequent derailment with additional dangers.

In line with all railroad safety trends, the number of main line track collisions have dropped significantly in the last 30 years (Figure 1.1). Not included in these statistics are collisions with highway vehicles or other obstructions on the tracks. A road crossing collision is usually not a serious event for the train, unless the vehicle is a heavy truck or, worse still, a truck carrying gasoline or other hazardous material.

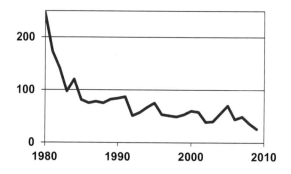

FIGURE 1.1. Number of main track collisions from 1980 to 2009.

Thirty percent of all collisions occur on the side of a train (most commonly when a siding joins the main track), 12% are head on, and 22% are in the rear. The most common collision is a raking collision, which takes place when derailed cars are dragged along the side of another train. (The derailed car "rakes" the second train.) A derailed car dragged into a bridge or other solid structure is also considered a raking collision.

Chapters 2 through 6 discuss collisions involving passenger trains and freight trains and the means to improve safety and prevent collisions.

Derailments

Main line derailments are 15 times more likely than collisions. Most derailments are relatively benign, and can be compared to a person walking down the street, tripping, getting back up, and continuing on her or his way. Unless derailed cars crash into houses, strike passenger trains, or release hazardous material into a neighborhood, derailments do not normally affect civilians. Generally speaking, a passenger train derailment is not as disastrous as a collision unless the train crashes into something solid at high speeds, derails off a mountain or bridge, or derails into water. Yet derailments have been responsible for some of the greatest train wrecks ever.[7]

Derailments are usually caused by equipment failures. Broken, settled, spread, shifted, or overturned rails account for about 50% of the equipment-related derailments. Poor train handling, incorrectly set track switches, unsecured cars on a hill, shifted loads, vandalism, or obstruc-

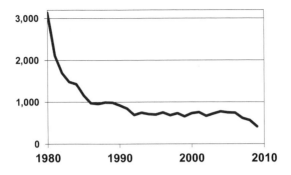

FIGURE 1.2. Number of main track derailments from 1980 to 2009.

tions on the track are among the human causes of derailments. Trains can jump off the tracks from truck-hunting or harmonic rock-off. Derailments can also be caused by flash floods, avalanches, rock slides, and high winds. Chapters 8 through 12 discuss various ways that a train can derail. Derailments have dropped dramatically since 1980 (Figure 1.2).

THE ACCIDENT INVESTIGATION PROCESS

Prior to the twentieth century, train accidents were investigated by the county coroner just as any other accidental death. In 1911, the ICC was given authority by Congress to investigate train accidents. The ICC could only enforce existing congressional laws and could not initiate safety rules on its own. In 1967, the FRA replaced the ICC. In 1970, Congress gave the FRA authority to "promote safety . . . and reduce railroad accidents" and the legal authority for enforcement.[8] The FRA formulated extensive safety regulations for track and equipment, operating rules, and civil and criminal penalties for violations.

In 1967, Congress also created the National Transportation Safety Board (NTSB) to investigate all transportation accidents, including rail, aviation, highway, marine, and pipelines, and to establish "probable cause" and make safety recommendations. In that capacity the NTSB may investigate FRA oversight and regulations. The NTSB handles the most serious or catastrophic accidents,[9] with special attention given to accidents involving passenger trains and hazardous materials.

On its own the FRA conducts about 100 formal accident investigations per year (including all accidents with fatalities). The NTSB can take over an FRA investigation if it chooses. All investigations include extensive engineering forensics and human factor studies. Accident investigations form the foundation for continuous incremental safety improvements and this book.

2

How Trains Crash,
Then and Now

Trains more or less crash the same way they did 100 years ago—albeit far less often. Trains still derail or fly off the tracks, usually because of equipment failure, and trains still collide, more often than not because of human error.

In spite of describing crash after crash, this book also emphasizes the many safety improvements that have historically occurred and continue to occur at an accelerating pace. Dramatic safety improvements are easily seen by comparing train wrecks "then and now."

MODERN COLLISION

The violence and destruction of a head-on collision is nearly incomprehensible.

On Friday, September 12, 2008, at 4:22 p.m. a Los Angeles Metrolink commuter train with 3 cars and 225 passengers drove past a stop light before colliding head on with a Union Pacific (UP) freight train. Twenty-five people were killed and 102 were hospitalized. It could have been a lot worse. The destroyed car on this pre–rush hour train was rated to carry nearly 5 times more passengers.

On the day of the accident, the Metrolink engineer sent and received

43 text messages while operating the train. The texting occurred with local teens—railroad enthusiasts who had adopted the Metrolink engineer as a mentor.

Signal lights along the rails control train movements just as traffic lights control car traffic. The signals tell the locomotive engineer to stop, go, slow, prepare to stop at the next signal, and so forth. Safety procedures required the Metrolink engineer to call out by radio to the conductor all signals encountered. The engineer called out the first of two warning signals (a flashing yellow light) at 4:17:45, indicating "prepare to stop at the second upcoming signal," but not the second warning signal (a solid yellow light passed at 4:18:41) or the final signal—"red for stop" passed at 4:21:56. This implies that the engineer was distracted, did not follow required safety rules, and probably missed the second and third lights completely. At 4:22:01, the engineer transmitted a completed text message just 22 seconds before the collision.[1]

The Metrolink train was signaled to stop and wait for the oncoming freight train to get off the main track and onto the siding. The collision occurred 1,761 feet (537 m) past the Metrolink's stop signal. The freight train, exiting a tunnel just 634 feet (193 m) from the collision, had a limited line of sight on the curved track (Figure 2.1),[2] and the engineer applied the emergency brakes just 2 seconds before the collision. The Metrolink engineer never braked.

Passengers on the left side of the train saw the approaching freight train around the curve. "My first thought was: I'm not seeing this," said one commuter on the Metrolink train. "I saw it coming," said another passenger in the second car. "There was no time to stop. The next thing I knew I was in a seat in front of me. It was horrible."[3]

Data recorders indicated that the Metrolink and Union Pacific trains were traveling at 43 and 41 mph (69 and 66 km/h). The speed limit on this track is 60 mph (97 km/h) for freight trains and 70 mph (113 km/h) for passenger trains. The speed limit on this curve, however, is reduced to 40 mph (64 km/h).

An eyewitness a few hundred feet away said, "I heard a huge crash, then I saw a fireball. I ran over there and there were people lying all over the hill."[4] Another eyewitness said, "People were climbing out of the side, bleeding, crying, screaming. It was like a war zone."[5]

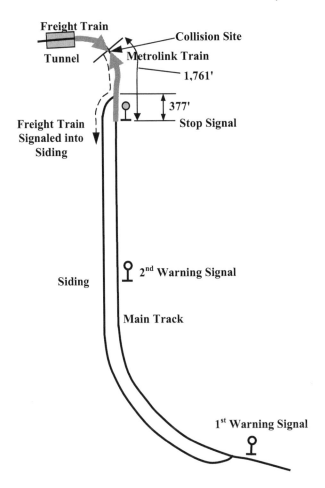

FIGURE 2.1. The collision site track and signals. The freight train had just exited a tunnel 634 feet (193 m) from the collision.

The 1,164-foot (355-m)-long UP train, including two 429,000-lb (194,590-kg) locomotives up front, weighed 1,522 tons (1,380 MT). The 313-foot (95-m)-long Metrolink train, weighing 308 tons (279 MT), consisted of one 270,000-lb (122,470-kg) locomotive and 3 passenger cars. The collision shoved the Metrolink train backward 76 feet (23 m), and derailed the 2 UP locomotives, the first 10 cars of the freight train, the Metrolink locomotive, and the first passenger car of the passenger train (Figure 2.2). The UP train crew, their cab intact, survived with serious injuries. The impact crushed the cab section of the Metrolink locomotive,

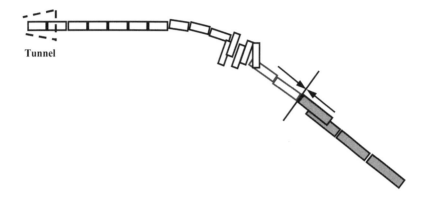

Tunnel

FIGURE 2.2. Collision diagram. Union Pacific freight train (white) and Metrolink train (gray). Note the Metrolink locomotive embedded in the first passenger car. *Adapted from California Public Utilities Commission.*

killing the engineer, and compressed the Metrolink locomotive from its original length of 58 feet down to 42 (17.7 to 12.8 m). The Metrolink locomotive was also shoved backward into the first passenger car. About 52 feet (15.8 m) of the 85-foot (25.9-m)-long passenger car was crushed. Twenty-two of the 24 passenger fatalities were in the first passenger car.[6] The crushed volume was generally not survivable. Just behind the crush zone had borderline survivability.

Swept ahead of the Metrolink locomotive crashing into the passenger car was a wall of collision debris consisting of a compressed 10-foot (3-m)-thick mass of seats, floor, ceiling panels, tables, posts, and body parts. The Metrolink locomotive's 2,200-gallon (8,328-l) fuel tank separated during impact, came to rest at the side of the track, leaked, and ignited. The fire was not a major factor in the rescue efforts—the mangled wreckage was.

The very first responders had trouble understanding the enormity of the disaster, but immediately saw the fire. The flames were confined to the Metrolink locomotive. With spilled fuel puddled an inch deep, the fire getting closer to the freight train's two 5,000-gallon (18,927-l) fuel tanks, and no signs of life in the first Metrolink passenger car, the first responders concentrated on evacuating the walking wounded from the second and third passenger cars (Figure 2.3). Some passengers were in shock and did not want to leave the train even after being told about the

FIGURE 2.3. Metrolink locomotive (foreground) embedded in the first
passenger car.

fire and potential fuel vapor explosion. Many others were in too much
pain to move.

Consider the confusion of the rescue workers. The derailed trains
were spread over many hundreds of feet. Those arriving on foot could
only see one side of the accident. A locomotive compressed 16 feet (4.8 m)
and embedded 52 feet (15.8 m) into a passenger car does not look like
anything within the realm of normal experience. Meanwhile, there was
total pandemonium as injured passengers milled about crying for help
and many more injured passengers lay inside the train wreckage. On-
lookers, good Samaritans, and media gathered. Billowing, thick black
smoke added to the confusion.

One fireman described the scene: "I did see some people kind of wandering around and they, everybody was just dazed and confused. They
were just kind of wandering around, . . . we told them, hey, go over to
the grass and sit down on the grass and wait there, and we're going to
get people there to take care of you. But we kind of went past that. Our

concern was the fire because we knew we had cars involved, and we wanted to get the fire knocked down."[7]

Another said, "I saw locomotives engulfed in flames, fire[fighters] in the process of putting it out, and my attention first was I saw numerous people, maybe a dozen, walking in various means, I don't know, delusioned, like they were zombies walking with various types of injuries with their hands out and saying help. And the first thing I did was run to their direction, to ease them, try to determine their injuries, assured them that, you know, paramedics would be coming. I boarded the first [passenger car] and started helping people clear off . . . that were injured . . . I went to the second [car], and there was people laid out trying to, you know, help the ones that could get off the train."[8]

The firemen's focus shifted to the fire, which was quickly extinguished once they set up hoses from the nearest hydrant. Shortly after, foam tenders arrived to apply foam on the fuel-soaked ground to prevent re-ignition. While putting out the flames, the firemen noticed the freight train engineer and conductor pounding on the window of their derailed locomotive. The train crew was trapped inside with smoke entering their cabin! The locomotive's video records the fireman fighting the fire at 4:42 p.m.[9] Firemen immediately attacked the windshield with axes. The 0.625-inch (15.8-mm)-thick, laminated glass, designed to successfully confront a 24-lb (10.8-kg) concrete block at 30 mph (48 km/h), would not budge. It took about 2.5 minutes to pry out a rubber seal and remove the window. Both injured crew were carried from their cab.

As the trickle of firemen, police, and county sheriffs turned into a torrent, they fell into a routine. The police and sheriffs assisted the firemen by evacuating and bandaging the injured, holding pressure on wounds, removing debris, and the like.[10] Police also secured the area, clearing out bystanders and media.[11] The firemen organized themselves into fire suppression, medical, extrication, and hazmat groups. Others positioned emergency lighting, food, water, and toilets for the long haul.

The medical group set up a triage area and a temporary morgue. Each passenger removed from the train was tagged green, yellow, or red (for light, moderate, or critical injuries) and moved to the corresponding triage area. The hazmat group tracked down the freight train's manifest

FIGURE 2.4. The Metrolink passenger car, showing stairs to three different levels.

(mostly fruits and vegetables), opened every car to verify contents, and continued to monitor the leaked fuel for explosive potential.

Because of limited access through the back door, the crush zone of the first passenger cars was blocked. The best access was the side door that was now on the top of the overturned car. Firemen laddered up and started peeling off the side wall (Figure 2.4). Rescuers began daisy-chaining debris—pieces of seat cushions, posts, tables, partitions, stairs, roof panels, flooring, and the like—out of the passenger car. Seats were unbolted for removal. Extricated victims were placed on cumbersome backboards and carried out by a line of police and firemen.[12] Some injured passengers had mutilated body parts and/or missing limbs. One of the first firefighters entering the car quickly estimated 2 or 3 could be extracted quickly, 6 or 7 were dead, and "about 8 or 10 were alive but weren't going to make it."[13]

The mangled, compressed wreckage was difficult to cut. Rescuers hesitated to use torches because of spilled fuel. One fire chief explained the problem: "I described it as if you've ever seen a bale of metal recycling, how tightly packed that is . . . so our normal urban search and rescue tools and techniques, you know, which really rely on being able to pry off something solid or cut something that's going to give you some sort of a sizable release that you can then make entry through, that impact area was packed in so tight that none of those traditional methods worked."[14]

Another fire chief added: "We cut, cut, cut, we spread, and it doesn't spread the way we thought. So we have to do another plan . . . in the process of spreading and cutting, then we start shoring. We start bringing in blocks in there because once we spread it with the spreaders, and then we remove a body and we take this tool out to go put it in another place, then the metal comes back . . . it was in a truly horrific environment. It was probably an area that was maybe 10′ by 10′ (3 by 3 m) that was full of body parts, blood and decapitated corpses."[15]

By around 8:00 p.m., all the injured had been transported to hospitals. The focus switched from search and rescue to body recovery mode around 1:00 a.m. The Metrolink engineer's body was removed from underneath his locomotive the following morning. The final victim was recovered around 2:00 p.m.

Extreme trauma patients quickly saturated the closest hospitals. Five helicopters flying 26 flights distributed the injured to numerous hospitals farther away. UCLA hospital treated 8 crash victims, with 5 undergoing immediate surgery. The following day 4 UCLA patients were in intensive care and in critical condition, 3 remained hospitalized in serious condition, and the 8th was released. The day after the accident, at least 86 passengers remained hospitalized, almost half in critical condition.

FOUR DAYS AFTER THE ACCIDENT, line-of-sight testing was done with two similar locomotives. The two test locomotives were placed nose-to-nose at the point of impact and then moved back incrementally. The distance between the trains when they could first see each other on the curve was 539 feet (164 m), about 4.4 seconds before impact. The rule of thumb used to calculate safe stopping distances conservatively assumes a 3-second delay before the human becomes aware of and responds to the signal. Obviously, if the engineer is expecting the signal and watching for it, he or she can respond faster.

Why Were the Trains on a Collision Course?

On average, 14 freight, 18 Metro, and 12 Amtrak trains travel this track every day. Most of this route is double tracked to accommodate two trains moving in opposite directions. Because there are three consecutive tunnels wide enough to carry only one track, this particular section of track is single track. Two trains moving in opposite directions must take turns moving on the single track.

Positive Train Control

Final passage of congressional legislature requiring Positive Train Control (PTC) occurred shortly after the Metrolink collision. PTC, a work in progress for many years and the subject of Chapter 6, is designed to safely stop a train if its operator misses a signal. This new safety system is being fast-tracked in southern California.

HISTORIC COLLISIONS

The worst train crash in American history occurred on July 9, 1918. Two passenger trains collided head on in Nashville, Tennessee, with 101 fatalities. This accident was very similar to the 2008 Los Angeles collision. Double tracks merged into a single track. One train failed to stop and wait for the other train to pass. The northbound train misunderstood its orders, failed to stop, proceeded on the single track, and collided with the southbound train. Both trains were estimated to be traveling about 50 mph (80.5 km/h). As horrific as the 2008 Los Angeles collision was, only one passenger car was destroyed. The collision in 1918 destroyed six wooden cars.

In 1918, the northbound train had two unoccupied baggage cars behind the locomotive; the southbound train (with most of the fatalities) had only one. If all cars behind both locomotives were occupied (as was the case in 2008 collision), many hundreds more could have died. If there was an unoccupied car behind the locomotive in the 2008 collision, the fatalities would have been reduced from 25 to perhaps 1 or 2.

Wood and metal differ in their responses to crash energy. Bend a strip of wood and it snaps. Crushed metal absorbs more crash energy. Metal will tear, but not as easily as wood. Certainly, steel cars will not shatter like wooden ones. The investigators of the 1918 collision correctly focused on the frailties of wooden cars.

WORSE STILL, wooden cars would often ignite after the crash. The worst train wreck in British history combined the frailty of wooden cars with the risk of fire. In 1915, a distracted signalman signaled a troop train into a stopped train. Immediately after the collision, an express train plowed into the troop train at high speed. The troop train was totally destroyed with its length reduced from 639 feet (195 m) before the collision to only 201 (61 m) after. Three cars of the express train telescoped.

Flying hot coals ignited the wooden cars. Gas lighting in the troop train added to the blaze. Pressurized containers filled with natural gas ruptured during the crash and subsequent fire. Fire burned out of control for almost 24 hours and destroyed 15 cars of the troop train, 4 cars of the express train, and 7 and 5 cars on 2 trains standing on adjacent tracks. A total of 226 people, most of whom were on the troop train, lost their lives.

Except for a train swept out to sea by a 2004 tsunami in Sri Lanka,[16] the worst train disaster ever was a runaway French train in 1917 that ended in a fiery inferno. Two trains left Italy with 1,000 French soldiers on board for Christmas leave from the Italian front. Because of a wartime shortage of locomotives, the two trains were combined into one at the French border. The locomotive was rated to pull 144 tons (131 MT). Only 3 of the 19 cars had airbrakes; the remainder had hand brakes. The experienced driver refused to leave the station and take the 546-ton (495-MT) train down a 3.3% grade,[17] an extremely dangerous hill even for today's trains fully equipped with modern brakes. During peacetime the driver's judgment would have prevailed. The engineer was threatened with wartime military discipline, court martial, and possible execution.

In spite of descending the hill with maximum braking effort at just 4 mph (6 km/h), the train began to run away. Sparks from the overheated brake blocks ignited the wooden floors. The train continued for a few miles before flying off the tracks at an estimated 75 mph (121 km/h) on a sharp curve near Modane, France. The wooden cars piled up against a retaining wall and burst into flames. There were 543 confirmed deaths and many more passengers unaccounted for. Runaway trains, an interesting lesson of gravity, still occur today for a variety of reasons discussed in Chapters 9 and 10.

OVERHEATED BRAKES IGNITING WOODEN FLOORS remained a problem for many years on freight trains. Obsolete journal-bearings can also overheat and ignite fires. Worse still, old-fashioned journal-bearings would spray lubricating oil on the bottom of the wooden floorboards.

In one 5-year period between 1968 and 1973 there were 237 car fires on just a single subdivision of the Southern Pacific railroad—14 from overheated bearings, 184 from overheated brakes, and 39 from other causes. One of those fires was particularly exciting.

On May 24, 1973, an overheated brake ignited a wooden boxcar packed with bombs. In fact, the 107-car train had 12 such cars filled with 2,600 500-lb (227-kg) bombs bound for Vietnam. The train was descending a 1% grade at 45 mph (72 km/h). The train crew was unaware of the initial explosion and of another that occurred 5 miles (8 km) later. The conductor in the caboose noticed a smoldering wooden railroad tie and grass fires and notified the engineer in the front of the train. The engineer looked back, saw gray smoke, and began slowing the train. The conductor and brakeman in the caboose saw a fire on the train, applied the emergency brakes, and jumped from the caboose. Another mile down the track a large explosion occurred.

A series of explosions occurred for the next 6.5 hours, destroying 12 munitions cars and about 2,100 bombs. The main blast hole created a 115-foot × 93-foot (35-m × 28-m) crater 25 feet (7.6 m) deep and scorched the desert for a 0.25-mile (402-m) radius. Unexploded bombs were thrown up to a mile. Windows in a home 5 miles (8 km) away broke. Fortunately, no one was killed in this sparsely populated part of the Arizona desert.

Fires still occur in box cars packed with flammable goods that ignite by other means. Modern bearings and brakes can still overheat, but the problem is significantly reduced. And today the bombs would be shipped in cars with metal floors.

WHAT SEEMS OBVIOUS TODAY was not so in the early part of the twentieth century. Steel was not the cheap and readily available commodity it is today—wood was. In fact, fearing electrocution from lightning, passengers were quite leery of riding in steel cars.[18] The railroad companies were even more suspicious of the expensive new technology.

The game changer for steel was the possibility of a tunnel fire. Fires in tunnels (even today) remain a special terror. A fire in the open spreads in all directions. In a tunnel the fire is concentrated and intensified. The fire, smoke, and carbon monoxide also make evacuation and rescue problematic. Such a fire occurred in 1903 in Paris, killing 84 in an underground station and tunnel. Many of the victims suffocated from carbon monoxide poisoning.

In 1903, plans were under way for constructing tunnels (under the East and Hudson Rivers) entering Penn Station in New York City. In 1910, steel cars were widely used for the first time in those tunnels.

Tunnels also create ventilation problems. It was not uncommon for the fireman of a steam locomotive to temporarily pass out in a long tunnel while shoveling coal into the firebox.

The worst train disaster in Italian history occurred in 1944, when an overloaded train stalled inside a tunnel near Balvano, Italy. Instead of safely backing the train out of the tunnel and down the steep hill, the engineer continued forward, stoking the steam boiler with low-quality wartime coal. The excess carbon monoxide poisoned 426 people—all of them stowaways on the 47-car freight train. Rescuers described victims as appearing to be gently asleep. The only survivors were in cars outside the tunnel.

Wooden cars were banned on the Pennsylvania Railroad, the richest company in the world, in 1928. In that same year, England (virtually bankrupt after World War I) had all wooden cars, 75% of them still lit by gas. By 1950, almost all passenger cars in America were steel.

Urban transit vehicles were slow to convert to steel. Being on isolated tracks, they were not at risk of being shattered by a heavy freight train. Also, because they were powered with electric power instead of engines with fuel tanks or beds of hot coals, the fire risk was reduced (but not completely eliminated). Nevertheless, there was a reported fire involving wooden cars on the Chicago Transit System as late as 1956.[19]

It is very difficult to assess the status of wooden passenger cars elsewhere in the world, but some information is available from documented train wrecks. In England wooden cars with metal frames were still in production in the early 1950s. In 1955, a British accident report (in which 11 died in a train fire) states that there were still 13,900 wooden cars with metal frames in England with steel replacement projected by 1963. In Spain a fire broke out on wooden cars and killed 30 in 1965, and 33 in wooden coaches died during a collision in 1970.

Surprisingly, even though wooden coaches, gas lights, pot belly stoves, and burning coals are no longer used, fire still remains a substantial danger. In Egypt in 2002, a fire started by a passenger's cooking stove destroyed 7 of 11 passenger cars, killing 373. Without contact with the back of the train (or fire extinguishers in each car), the train continued at 70 mph (112 km/h) for 4 hours, fanning the flames. The overcrowded train (with wooden seats) had passengers crouching in the aisles and lying in the overhead bins. Because of more concern about fare dodgers than safe egress, the windows had steel bars.

The very latest materials are quite fire resistant. In one test an 85,000 BTU/hour (25,000 watts) gas flame was used to ignite curtains and seats. (All 4 burners combined on a typical gas stovetop create 35,000 BTU/hour [10,200 watts] of flame.) The fire did not spread beyond the immediate flame contact. As the gap grew between the flames and burned materials, the fire extinguished itself. Even fire-resistant material will spread a fire if the fire is large enough.[20]

Tunnel Fires

Tunnels remain scary. In 2000, 155 passengers on an Austrian ski resort train died inside a tunnel. Without an engine or fuel, the cable-pulled train built with fire-resistant materials was considered fireproof. Proving anything is possible, flammable brake fluid leaked onto a faulty electric heater. The 12 survivors evacuated downhill away from the carbon monoxide fumes.

Fuel sprayed from a fuel tank after a collision remains a problem. In 1999, an Amtrak train struck a truck at a crossing. Leaking fuel ignited a sleeping car. The coroner ruled 5 had been killed by fire and 6 by trauma.

Fire disasters also reappear in the modern era with the transportation of tanker cars filled with pressurized petroleum gases. (See "Bearing Failure Triggers Chain of Events" in Chapter 8.)

Steam Locomotives

In yet another obsolete terror, the energy of pressurized steam could reappear as an exploding boiler. On March 18, 1912, a locomotive exploded in a repair shop in San Antonio, Texas, killing 26. A 16,000-lb (7,257-kg) section was blown 1,200 feet (366 m). Another 900-lb (408-kg) part blasted more than 2,200 feet (670 m), ripping out the side of a house. The attached oiler tender was thrown 150 feet (46 m), and sprayed hundreds of gallons of fuel oil that ignited numerous fires.

Exploding boilers were usually only a danger to the train crew, but not always. In 1934, a boiler exploded out of a locomotive and blasted 50 feet (15 m) into the air before landing on a wooden passenger car, killing 17. A similar accident occurred in 1935, when a steam locomotive pulling 4 wooden cars packed with 300 miners exploded. Two members of the locomotive crew were tossed 100 yards (91 m). The boiler blasted straight up before crashing into the first wooden car. Two crew and 16 miners were killed. (See "Boiling Propane" in Chapter 8 for a description of boiler explosion energy.)

In other accidents passengers have been scalded to death. In 1925, a locomotive derailed onto its side in New Jersey. The second coach tipped over on top of the engine. Most of the 50 fatalities were due to scalding after the boiler exploded. A similar accident occurred in 1943, when a freight train boiler exploded after colliding with a passenger train. Twenty-six of the 27 victims were scalded to death.

The dangers of pressurized steam were eliminated with the adoption of the more efficient diesel engine. Diesel locomotives, first used in the United States in luxury passenger trains in the 1930s, rapidly replaced steam locomotives in the United States by the mid-1950s.

The diesel is an internal combustion engine. A steam engine uses external combustion—a pile of coal burning outside the piston. Internal combustion captures more heat energy and operates with a higher thermal efficiency.[21] The best steam engine had a thermal efficiency of 12%; that of diesels can be as high as 35%. Also, the steam engine had to carry around its working fluid—water. The diesel's working fluid is air, energized by exploding fuel mist. A diesel engine is also far less labor intensive.

Moderate-Speed Passenger Train Collisions

The crowded 12-car New York City rush hour commuter Train 780 left Manhattan at 6:09 p.m. on November 22, 1950, with more than 1,000 passengers. Another identical train, Train 174, with an estimated 1,200 passengers left the same station just 4 minutes later. Typical with commuter trains at rush hour, many passengers were standing.

Responding to a signal, Train 780 slowed from 30 to 15 mph (48 to 24 km/h). The train continued to slow and unexpectedly stopped at 6:26 p.m. Thinking the brakes were not releasing properly, the engineer switched from electro-pneumatic braking to conventional airbrakes.[1] Still trying to release the brakes, the engineer applied (and released) the conventional airbrakes and then the emergency brakes. The signals automatically signaled the train behind to stop.

Just 3 minutes after Train 780 stopped, Train 174 accidentally drove past its stop signal and collided into the rear of Train 780 at 30 mph (48 km/h). Seventy-eight passengers and crew died. The first car of Train 174 was destroyed (resulting in most of the fatalities), as was the last car of Train 780. Demonstrating how concentrated the damage (and crash energy) was, only the last car of Train 780 derailed. All the other cars remained on the tracks and attached to their respective trains.

MULTIPLE-UNIT TRAINS

Commuter trains, making very short trips with frequent stops, want to turn around and repeat their route as soon as possible. For that reason commuter trains often operate in a push-pull mode to avoid turning around. In pull mode a locomotive in front pulls the train. In push mode the locomotive in back pushes the train forward.

Of course, the train crew must be able to see where they are going. During push mode the crew controls the train from a special cab car at the opposite end of the train. A cab car has a set of controls electrically connected to the locomotive in the rear. About 65% of U.S. commuter trains operate in push-pull mode with a cab car. In another variation commuter trains operate multiple-unit, or MU, cars that are self-propelled; these trains do not have a traditional locomotive.

Commuter trains that frequently stop and start need large forces to provide the acceleration. If too much torque is placed on one wheel, however, it spins. The solution is to use many smaller motors distributed across the train with smaller torque on each driven wheel—hence, MU cars.

Cab cars for MU trains are structurally less significant than a locomotive with a heavy engine. MU cab cars, with seats for passengers, place the passengers closer to the point of impact. The 1950 New York City collision was between two MU trains without locomotives.

Colliding commuter trains is the worst possible scenario. Commuter trains are likely to be crowded with standing passengers. Intercity passenger cars will often have an empty car behind the locomotive filled with baggage or a mostly empty diner car. Intercity train seats are padded for comfort (and more safety). Most commuter cars have hard plastic seats. Locomotive-led trains separate passengers from the primary collision. In the same year as the 1950 MU train collision a locomotive-led passenger train traveling at 40 mph (64 km/h) struck the back of a stopped freight train with no fatalities.

The most dangerous crash scenario for passengers is loss of livable volume due to crushing, telescoping, overriding, or some combination of the three.

TELESCOPING

Telescoping occurs when the underframe of one car overrides a second car and slices through the weaker side walls. A car's underframe is much stronger than the side walls because the underframe must support the car body, as well as withstand the forces that pull the train and the impact of coupling.

The telescoping car slides into the telescoped car (the car with greatest injury), like closing a collapsible telescope, crushing everything in its path. A classic example occurred at Mud Run in 1888, when a train missed its signal to stop and rammed the rear of a stopped train. The locomotive of the wayward train plowed through 20 feet of the last car of the stopped train. The last car was then shoved into the next car, telescoping half its length (Figure 3.1). The last two cars contained 200 passengers, 66 were killed.

FIGURE 3.1. After a collision, the last car of a stationary train telescopes half of the next car. *Mud Run, 1888; Library of Congress.*

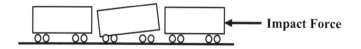

Impact Force

FIGURE 3.2. Car-pitching leads to override. Two opportunities for overriding are shown: the car on the left can override the middle car, and the middle car can override the car on the right.

Any mismatch in heights of the underframes during the collision can cause overriding. Underframe mismatch occurs during a collision for a variety of reasons, including track irregularities, difference in wear between wheels, or unequal braking during the accident. (If the brakes on one set of wheels grab more than the other set, the car will rock on its spring-mounted suspension.) Crushing during the collision can also deflect the underframe of one car above that of the other.

Steel cars can also telescope, though not as easily wooden ones. The stronger steel cuts both ways, literally. The strength of steel makes it more resistant to telescoping, but it also enables it to slice through the side walls of telescoped cars. The trick is to design cars less likely to telescope. If colliding steel cars do not telescope, the outcome will be superior to wooden car collisions.

In the 1950 New York City collision the last car of stopped Train 780 deflected upward and telescoped the first car of the colliding train for an undisclosed length. Both cars were 65 feet (19.8 m) long and weighed 114,000 lbs (51,709 kg).

Even cars in the middle of a train can telescope. Depending on which underframe is higher, Figure 3.2 shows two possible override scenarios. The car on the left is lined up to override the middle car, and the middle car could override the car on the right.

Apparently, this pitching motion occurred during a December 31, 1944, collision in Utah. The first train, a locomotive and 18 passenger cars, slowed to 8 mph (12.8 km/h) behind a freight train experiencing an overheated bearing. The second train, consisting of 20 passenger cars and a locomotive, was traveling at 65 mph (105 km/h) and braked just 12 seconds before colliding into the rear of the first train at an estimated speed of 50 mph (80 km/h). The 13th car of the first train telescoped the 12th car about 45 feet (13.7 m), the 16th car telescoped the 15th car about

40 feet (12.2 m), and the locomotive of the second train telescoped the rear car of the first train about 13 feet (12.2 m). All cars were constructed of steel. There were 50 fatalities and 81 injured.

Ordinary couplings slide apart vertically. Tightlock couplers, first used in the 1930s and made mandatory on passenger trains in 1956, resist vertical uncoupling with at least 100,000 lbs (444,822 N) of force during accidents. This provides resistance to the underframe lifting, the motion required for overriding and telescoping. Keeping the cars attached also reduces the possibility of cars impacting each other (or other solid objects) during accidents. The benefit of tightlock couplers was illustrated in a 1986 derailment of an antique train. Fourteen of the 23 cars derailed on bad track. The 10 derailed cars with tightlock couplers remained upright. The 4 cars without tightlock couplers were destroyed by overturning or jackknifing.

Poor design and bad manufacturing can also trigger telescoping, as illustrated in a 1972 Chicago collision between two multiple-unit commuter trains. During morning rush hour on October 30, 1972, Train 416, consisting of 4 brand-new bi-level Highliner self-propelled cars, overshot the station by 600 feet (183 m). In violation of operating rules (and common sense) Train 416 backed up and was struck by Train 720, which was traveling at an estimated 44–50 mph (71–80 km/h). Train 720 consisted of six single-level MU cars built in 1926.

Although the new Highliner cars were structurally superior, the frame of the older car overrode that of the Highliner.[2] The first car of Train 720 crushed about 10 feet (3 m) as it lifted up and sliced through the weakest part of the Highliner car at floor level, telescoping it for about 35 feet (10.6 m). All of the crash energy was absorbed by the telescoping cars, and none of the cars derailed. Most of the 45 fatalities occurred in the telescoped car of the train backing up. If the accident had not happened 200 yards (183 m) from a hospital, the fatalities would have been much higher.

The wheels, bearings, and axles of train cars are mounted within trucks. Freight cars lift off their trucks for ease of maintenance. Passenger car trucks are connected to the car body for safety. The added weight of the trucks makes lifting the underframe harder. If the trucks remain attached, it is harder for one car to telescope another. Today the connection between the trucks and a modern passenger car body must resist a 250,000-lb (1,112,000-N) force before shearing off. The trucks of the 1926 cars were not attached and came off during the accident. Also, each of the Highliner's collision posts (two I beams attached by welding) was supposed to shear off after 300,000 lbs (1,334,466 N) of crash force. The welds were not fully fused, with only 25% properly connected. The operational details of this crash (signals, stopping distances, line of sight, etc.) are presented at the beginning of Chapter 5.

Another head-on collision of multiple-unit commuter trains occurred in Chicago in 1984 with speeds of 30 and 15 mph (48 and 24 km/h). The vestibules of the lead cars of both trains were crushed and derailed. Both control cab compartments were crushed, but both crews got out in time. Of the 87 people injured, 9 were admitted to hospitals with fractures and concussions. The two colliding cars coupled together, remained coupled, and eliminated the possibility of telescoping.

OVERRIDING

When the underframe of one car rides up or overrides the underframe of the car being hit in a collision, telescoping can result. Overriding refers to a different impact mechanism not associated with telescoping. The crushed ends can form a ramp that catapults one car on top of the other. (Often the terms *overriding* and *telescoping* are used interchangeably. Some may call the overriding and ramping mechanism telescoping.) Ramping during collision is complicated, with many variations, depending on the contact misalignments and the sequence in which individual components fold, crush, collapse, and/or fracture. Often the car with the least weight bounces up and overrides the heavier car during a collision.

IN 1981, a freight train with a 247,000-lb (112,000-kg) locomotive pulling 4 cars at 12 mph (19 km/h) collided head on with a commuter train led by a 85,000-lb (38,555-kg) cab car moving at 19 mph (30.6 km/h). Surprisingly, the cab car got the better end of the deal. The cab car underframe crushed, peeled away up to the wheels, and folded, forming a ramp. The ramp caused the cab car to override the freight train locomotive in a manner very similar to what is shown in Figure 3.3. Three crew

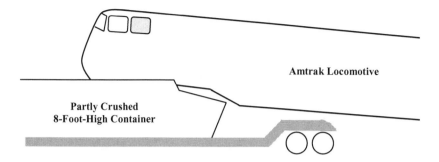

FIGURE 3.3. An Amtrak locomotive overrides a flatbed freight car with a container.

were killed in the freight train locomotive. The commuter train operator was killed after being thrown from the cab car. Twenty-eight passengers were injured.

This accident is very similar to a 2002 crash test. A 75,000-lb (34,000-kg) cab car–led train with 4 cars and a trailing locomotive moving at 30 mph (48 km/h) struck a stationary 244,000-lb (110,676-kg) locomotive with 2 attached hopper cars. Both trains weighed 635,000 lbs (288,000 kg). Within 0.2 second after impact, 18 feet (5.5 m) of cab car underframe crushed. The cab car ramped and rose when its underframe crushed. After 0.7 second and 22 feet (6.7 m) of underframe crush, the cab car rode up on the roof of the locomotive with a 9-degree tilt. About 2 seconds after impact, the 2 trains were moving together at 15 mph (24 km/h). The cab car–led train slowed faster and fell off of the locomotive 6.5 seconds after impact.

In the cab car seating for the operator and 47 simulated passengers was crushed during the test. The crash dummies away from the crush zone "survived." In this case the locomotive was relatively undamaged, with only 1 foot (0.3 m) of crush near the windshield. Very little damage occurred to any other cars.

The kinetic energy of the 30-mph (48-km/h) train was a little more than 19 million foot lbs (25,760 kJ), about the same energy required to lift the 635,000-lb (288,000-kg) train about 30 feet (9.1 m).[3] The researchers determined that half of the energy was dissipated by crushing metal; the rest was dissipated by sliding the freight train (with its brakes on) backward. The cab car crush force was only slightly less than the force required to shear off a component housing on the locomotive. If the housing had sheared off before the cab car crushed, the locomotive might have overrode the cab car.

During a 2007 collision, an Amtrak train traveling at 36 mph (58 km/h) struck the back of a standing freight train. The Amtrak locomotive overrode the freight car, as shown in Figure 3.3. Two of the 66 passengers were hospitalized with serious injuries; the crew received only minor injuries. The Amtrak locomotive, with an underframe reinforced to pull a train (and support the engine), had limited bottom crush.

During a 1979 collision, overriding of two different locomotives apparently absorbed most of the crash energy and protected the passengers when an Amtrak train collided at 58 mph (93 km/h) with a stopped freight train. A track switch incorrectly directed the Amtrak train (1 locomotive and 5 coach cars) into a stopped freight train (3 locomotives and 40 cars). Upon impact, the Amtrak locomotive pushed the first freight train locomotive 34 feet (10.3 m) under the second locomotive, killing 2 on the freight train. The Amtrak locomotive then struck and overrode the second freight train locomotive. The Amtrak locomotive and the first car overturned. Thirty-eight of the 210 passengers were injured. The two colliding locomotives were destroyed.

LATERAL BUCKLING

The destruction of the livable volume by crushing, overriding, and/or telescoping is always the most dangerous accident scenario. Ideally, a train remains in line with all cars connected after a collision or a derailment. A train coming apart with cars flying off the tracks is the second most dangerous accident situation. Of course, if the train comes apart and cars crash into solid objects (e.g., a bridge abutment) or into each other at high speeds, the passenger volume is at risk.

During a collision or derailment, the kinetic energy has to go somewhere. Some is used up by the brakes trying to stop the train; some is used up by crushing metal. Much is used up by changing the direction of the cars' intended motion. The remaining kinetic energy of derailed cars is eventually dissipated by the cars sliding on the ground.

Newton's Law $f = m \times a$ can still be used to explain the motion of a train after it leaves the track. Things become more complicated in that the

FIGURE 3.4. Saw-tooth buckling schematic.

forces can be occurring in three spatial directions: along the X, Y, and Z axes (the same directions can be described as north, east, and up). The formula $f = m \times a$ must be applied three times, once for each direction.

There is also a rotational version of Newton's Law. Just as all masses resist acceleration (inertia, the resistance of mass to acceleration, is discussed in Chapter 7), they also resist rotational acceleration. The cars can move off the tracks with six types of motion, moving along the X, Y, or Z axes and/or rotating about the same three axes. It is difficult to imagine, but throw a tire off a 10th-floor balcony while spinning and flipping it over. All the motion of the tire, even after it bounces, slides, rotates, and spins, can be described as the summation of motion along the three axes and rotations about the same three axes. Given all the potential motions that can occur during an accident, there is surprisingly a strong bias toward one type of initial motion.

The tightlock couplers between cars form a rigid link during collisions. If the cars remain attached (a very good outcome that limits overriding and telescoping), they tend to saw-tooth buckle, as shown in Figure 3.4.

One 1984 collision possibly demonstrates the limits of tightlock couplers to keep a train together after a collision. Both passenger trains (one with 5 cars, the other with 7) were traveling at 30 mph (48 km/h) for an approach speed of 60 mph (97 km/h). The locomotive and first 4 cars of each train derailed, but all cars remained upright and inline on a death-defying 80-foot (24.4-m)-high bridge. Four of the derailed cars saw-tooth buckled. The 200,000-lb (90,700-kg) locomotives of both trains were shoved into their first coach cars. The first car behind each locomotive was crushed to the passenger compartment bulkhead. Crush damage on the ends of several other cars occurred in the vestibules. One passenger standing in a vestibule was crushed to death. One hundred and forty passengers and crew were injured, 10 seriously. About 2 feet (0.6 m) of crush damage in each locomotive helped absorb the crash energy.

AFTER SAW-TOOTH BUCKLING BEGINS, the large crushing force during a collision exerts a sideways force on adjacent cars. With increased crash speed, the saw-tooth buckle can grow into a large displacement lateral buckle with many cars derailing together on the same side of the tracks. In another configuration the train collapses like an accordion with many jackknifed cars. The zigzag buckling may increase until the cars have side-to-side impacts.

Longer trains traveling at higher speeds are more likely to have larger lateral deflections during an accident. Also, most collisions occur on curves where the line of sight is limited. Trains colliding on sharper curves have a greater initial offset that is more likely to induce a saw-tooth buckle. Attempting to sort out the variables, computer simulation compared lateral buckling of a six-car train and a nine-car train. Both simulated collisions occurred with an approach speed of 60 mph (97 km/h) into an identical train on a 2,200-foot (670-m) radius curve. The 9-car train laterally displaced 12 feet (3.6 m), the shorter train 5 feet (1.5 m).

A violent lateral deflection occurred in 1999. An Amtrak train with 2 locomotives and 12 cars derailed at 79 mph (127 km/h) after colliding with a truck at a road crossing. The truck, loaded with steel rebars, weighed nearly 75,000 lbs (34,000 kg). All 11 fatalities occurred in the 3rd passenger car, which bent 56 degrees when its side struck a locomotive. The front side of that car also crushed 6 feet (1.8 m) from additional car impacts. Leaked fuel ignited, and the coroner ruled that 5 of the 11 deaths were fire related.

Perhaps the worst example of zigzag buckling with side-to-side car impact occurred during a train wreck on June 3, 1998, in Germany. Part of a broken wheel snagged a track switch, throwing part of the train onto a parallel track at 120 mph (193 km/h). The third car slammed into a bridge, which collapsed on top of Cars 5 and 6. Five cars then slammed into a zigzag pile compressed to barely more than one car length. Of the 293 passengers and crew, 101 were killed.

The triggering event of the German disaster was metal fatigue of a steel tire. The separate steel tire ring was mounted on top of a rubber ring used to damp vibrations. This tire design had previously been used only on low-speed trams. This novel approach required aerospace quality design, testing, and inspection, none of which occurred. The conductor had the opportunity to safely stop the damaged train after a passenger complained he was nearly speared by the broken steel tire coming through the floor. The conductor, following procedure, chose to investigate before applying the emergency brakes. While walking to the damaged car, the conductor's sense of urgency should have increased when the cars started swaying. This German crash remains the world's worst modern high-speed train wreck.

OTHER LOCOMOTIVE–CAB CAR COLLISIONS

The saw-tooth lateral offset led to unexpected results during a collision in 2002. While traveling at 49 mph (79 km/h), a 5,700-ton (5,170-MT) freight train with 3 locomotives missed a stop signal. After 33 seconds of emergency braking, the freight train struck the cab car of a stopped Metrolink commuter train head on at 23 mph (37 km/h) on straight track. In spite of having its emergency brakes on, the 300-ton (272-MT) commuter train was shoved back more than 240 feet (73 m). Unfortunately, there were 2 fatalities and 22 serious injuries during this relatively low-speed collision. The deaths and injuries occurred when the passengers' forward motion was stopped by table tops in front of their seats. After this accident, collapsible table tops were developed and are now available.

Surprisingly, the worst damage and injuries were not at the point of impact, but at the back end of the struck cab car, which bent about 30 degrees and buckled laterally almost 8 feet (2.4 m) and vertically about 3 feet (0.9 m). The structural failure and buckle occurred about 15 feet (4.5 m) from the back end, where a section of the cab car broke off and telescoped itself for about 6 feet (1.8 m) (Figure 3.5). The couplers between the first and second passenger cars created a saw-tooth buckle and lateral loading that caused the damage.

Single-level passenger cars have a straight underframe with uniform strength. The underframe of a multilevel Metrolink car dips to accommodate another level (see Figure 2.4 in Chapter 2). The bend at the change in levels is a structural weak point that collapsed during the collision and created unexpected damage and injury away from the impact surfaces.

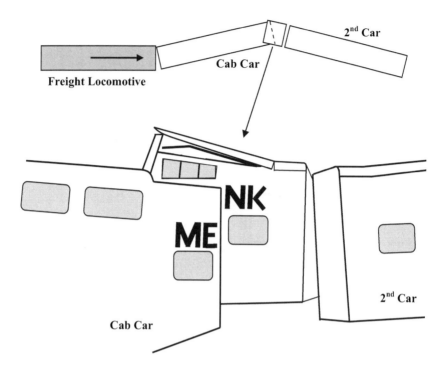

FIGURE 3.5. A 15-foot (4.6-m) section of a cab car telescopes itself during a 2002 collision at 23 mph (37 km/h). Note misalignment and missing letters in METROLINK.

IN 2005, a cab car–led commuter train traveling at 74 mph (119 km/h) saw an SUV stopped on the tracks ahead. The train slowed to 62 mph (100 km/h) before striking the Jeep Cherokee. The driver of the SUV had parked his vehicle not at a road crossing but instead directly on the tracks. If the SUV had stopped at a road crossing, the bottom of the tires would have been even with the top of the rails, and, most likely the train would have swept the vehicle safely aside. Because the Jeep was parked on the tracks, the tires were below the rails, making it easier for solid parts of the SUV to wedge under the cab car and derail the train.

After impact the front truck of the cab car re-railed onto a siding and struck a parked freight locomotive at 49 mph (79 km/h). The collision with the locomotive crushed the cab car 26 feet (7.9 m), killing 8. After colliding with the freight train, the second passenger car swung out and struck another commuter train moving in the opposite direction on adjacent tracks. Three more died on the second train. Eyewitnesses reported that the driver of the SUV got out and doused his vehicle with gasoline. He was sentenced to 11 consecutive life sentences. In his defense the driver testified he wanted to commit suicide but changed his mind.

CARS THAT DON'T CRUSH

Why can't we design passenger cars that don't crush, telescope, or override? We can, but it only creates other problems.

Mathematically, there are two extremes of collisions: the perfectly plastic collision, in which the colliding masses stick together like wads of bubblegum and cancel each other's momentum, and the elastic collision, where the two objects bounce away like billiard balls without any permanent deformation.

Consider a ball dropped onto a rigid floor. If the ball is made of bubblegum, all the kinetic energy of motion is absorbed by deforming; there is no bounce-back. For a perfectly elastic billiard ball collision, the billiard ball absorbs all the kinetic energy during the elastic collision and gives it back upon rebound. If the billiard ball is dropped from 30 feet (9.1 m), it strikes the floor at 44 ft/sec, or approximately 30 mph (48 km/h). In a perfectly elastic collision, identical to cars with zero crush, the ball bounces back at 30 mph and reaches its original height of 30 feet.

Now consider a perfectly elastic car that crashes into a rigid wall elastically (moving to the right) with no crush at 30 mph. An instant after impact, however, the car rebounds in the opposite direction at 30 mph, moving to the left. Unfortunately, the passenger sitting inside is still traveling 30 mph to the right. The net effect is that the passenger crashes into the inside of the car at 60 mph (97 km/h). This problem can be solved at the expense of passenger comfort with the use of crash helmets and a military-style lap and shoulder harness. The most comfortable way to travel is converted into the most annoying.

Not even military gear will solve the problem of a high-speed train collision. No one has ever suggested that it is safe to crash a car into a concrete wall at 100 mph (161 km/h) with a harness, crash helmet, and airbag. Newton's Law applies to your internal organs too. A body in motion, including your heart, tends to stay in motion. No matter how well your body is strapped in, eventually your aorta will tear as your heart continues forward. The rule of thumb for an aortic tear during a forward collision is a deceleration of about 50 Gs, or 50 times the acceleration of normal gravity.[4]

Consider a train traveling at 60 mph (97 km/h), or 88 ft/sec, a slow speed for a passenger train. It has an elastic collision in 0.1 second. (If such a thing were possible, the crash pulse for an elastic collision between two trains must be shorter and similar in duration to two colliding billiard balls.) If the passenger car is moving at 88 ft/sec and suddenly elastically rebounds to 88 ft/sec in the opposite direction within 0.1 second, the average acceleration is

$$2 \times 88 \text{ ft/sec divided by } 0.1 \text{ second} = 1{,}760 \text{ ft/sec}^2 \ (536 \text{ m/s}^2),$$

or about 55 times greater than the normal acceleration from gravity of 32 ft/sec² (9.8 m/s²). It doesn't matter how well you are attached to the seat; your aorta has torn during this crush-free collision.

There are other more practical problems for a crush-free design. Consider a World War II–style Sherman tank with a 350-horsepower (261-kW) engine that transports 3 passengers at 30 mph (48 km/h). Assume this tank can withstand a 60-mph (97-km/h) collision without crush. To travel twice as fast at 60 mph it will take 2 times more horsepower, or about 233 horsepower per passenger. Amtrak's 150-mph (241-km/h) Acela transports its passengers in style with about one-fifth the horsepower. Our hypothetical crush-free passenger car will cost at least 5 times more to build, uses 5 times more fuel, only moves at 60 mph, and still tears your aorta in a high-speed collision!

Instead of a rigid, billiard ball design, one alternative is having the train compress elastically many feet to absorb the energy like a large, helical spring. Assuming the cars remain in line during the collision (a very big assumption), a train moving at 120 mph (194 km/h) and decelerating a comfortable 5 times normal gravity would have to compress 95 feet (29 m). No one can sit in the volume space being compressed. For a train structure to elastically compress 95 feet it has to be many times longer, say, 300 feet (91 m) long. Three hundred feet of compressible structure would be needed on the front and back of each train being protected. This is not being suggested as a practical solution. Another option is to design cars that deform like wads of bubblegum to absorb energy. As we will see, this is in fact the modern approach.

STRENGTH OF PASSENGER CARS

All passenger cars that travel on the national rail network are required to sustain a static load of 800,000 lbs (3,560 kN) of force along the line of the couplers without any permanent deformation. This requirement dates back to the early part of the twentieth century.

Beginning in the late nineteenth century, postmen sorted mail overnight in postal cars attached to freight trains. As freight trains became longer and heavier in the early twentieth century, the flimsy postal cars would crush during an otherwise ordinary freight train accident. This led to standards on compressive strength of postal cars. The standards were quickly adopted for passenger cars and are still required today.

In the 1930s, lightweight, high-speed trains were introduced. These cars only had a compressive strength of 400,000 lbs (1,780 kN). The traditional cars with a compressive strength of 800,000 lbs (3,559 kN) were commonly mixed with the new 400,000-lb cars until an accident in 1938 in which a standing 11-car locomotive-led passenger train was struck head on by a 14-car locomotive-led passenger train moving east at 30–40 mph (48–64 km/h). Three crew died in the two locomotives. The only passenger cars destroyed were the first lightweight cars on both trains, the fourth car in the stationary train, and the third car in the moving train. The lightweight third car of the moving train had 8 passenger fatalities when it was telescoped 18 feet (5.5 m) by the car in front. Not mixing high- and low-strength passenger cars became standard practice moving forward. Today American passenger cars are still required to sustain a compressive force of 800,000 lbs at the couplers without any permanent deformation.

While 800,000 lbs (3,560 kN) of compression force sounds like a lot, the crushing forces during a train wreck are much higher. In 1999, a typical single-level American passenger car was crashed into a solid wall at 35 mph (56 km/h).[5] The car crushed 66 inches (1.68 m). The crash loads greatly exceeded the 800,000-lb strength requirement, as shown

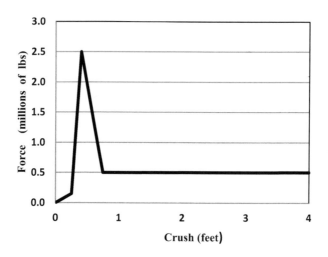

FIGURE 3.6. Idealized crash force versus crush of a typical American single-level passenger car.

in Figure 3.6. Higher-speed impacts always create higher crash forces. Consider gently placing a hammer on your head versus a high-speed hammer impact.

With just a few inches of crush, the underframe begins to buckle or collapse. Imagine pressing on a wooden yardstick. Initially, the yardstick is straight and rigidly supports its maximum load. Above some critical load, the yardstick begins to buckle and the load it can support drops off dramatically. The peak load in Figure 3.6 drops rapidly after the underframe buckles.

Once the underframe of the first car buckles, it can only transmit a load of about 500,000 lbs (2,224 kN), not enough to crush the cars behind it. This explains why most of the crash damage is usually focused at the colliding surfaces.

A follow-up test crashed two coupled passenger cars into a solid wall. Attempting to match the kinetic energy of the single-car crash test, the speed was lowered to about 26 mph (42 km/h). The lead car crushed about 6 feet (1.8 m) during this second test. The two cars remained coupled, but buckled in a saw-tooth pattern with a lateral displacement of 30 inches (0.76 m) during the collision.

THE SYMMETRY OF CRASHING INTO A WALL

When a train crashes into a wall at 35 mph (56 km/h), the train exerts a force on the wall. Per Newton's Law, the wall exerts an equal (but opposite) force on the train. An equivalent condition is two identical trains moving in opposite directions at 35 mph and colliding. Each train will exert an equal and opposite force on the other train. Therefore, crashing into a solid wall at 35 mph is equivalent to two identical trains crashing head on with a combined approach speed of 70 mph, or any combination of speed that adds up to 70 mph, for example, 70 and 0, 60 and 10, 50 and 20, and so forth.

The components of two identical cars in a perfectly symmetric crash at identical speeds, however, will not crush, fracture, fold, or collapse at the same time or in the same sequence. Two apparently identical parts on two different cars will not fail identically during a collision because of variations in material strength, dimensions, and/or assembly alignments. For the same reason, repeated crash tests with identical equipment will have different results.

For the same reason as well, computer simulations of collisions have limited accuracy. The components have too many choices on how they fail and in what sequence. The computer model has to be set up to focus on individual failure mechanisms. For example, overriding is very sensitive to misalignments. With an artificially introduced misalignment, the computer simulation is forced to focus on overriding. Repeated simulation of overriding results in a design that better resists overriding. Nevertheless, a computer model supplemented with crash-testing will give excellent results. Computer simulation is now considered to be within the accuracy of the repeatability of crash-testing.

HIGH-SPEED TRAINS

European freight cars have always been smaller than their American counterparts. Because of the shorter distances in more densely populated Europe, the Europeans never had the economic incentive to develop long, heavy freight trains. Also, tracks laid out 100 years ago frequently pass near delicate historic structures that cannot withstand vibrations from a heavy freight train. Today European freight cars are limited to 190,000 lbs (86,182 kg) compared to 286,000 lbs (129,727 kg) for American ones. Because of this fundamental difference, there never was a need in Europe to develop postal cars (and passenger cars) with an axial strength of 800,000 lbs (3,558 kN). In fact, today European passenger cars have an axial strength of only 450,000 lbs (2,000 kN).

After World War II, devastated Europe (and Japan) could not afford 100 million automobiles. Europeans were going to travel by train or not at all. Europe built national rail systems. Americans, who loved their automobiles, built a national highway system. Because of those different paths, Europe (and Japan) has become dominant in high-speed train technology.

When Americans finally wanted to build the 150-mph (241-km/h) Acela at the end of the twentieth century, they were forced to buy European technology. Amtrak could not simply buy an existing European design, however. As already explained, cars that crush at 450,000 lbs (2,000 kN) cannot be safely operated with cars that crush at 800,000 lbs (3,558 kN). In a collision the stronger car destroys the weaker car. To make the point obvious, would anyone feel safe in a car that weighs 1,500 lbs (680 kg) when all the other cars weigh 3,000 lbs (1,360 kg)? The European suppliers had to redesign their established technology. Many design errors and delays resulted.

The 800,000-lb (3,558-kN) axial strength requirement also led to other engineering problems. As we will see shortly, the modern approach designs a passenger car with a controlled crush to absorb crash energy and protect the occupied passenger space. Design for strength and design for crush are two contradictory requirements.

THE WASHINGTON METRO

The Washington Metro, like subway systems elsewhere in the United States, does not share tracks with the national rail network. Metro trains are not in danger of colliding with freight trains or other passenger trains. Isolated from all road crossings, the Metro cannot collide with road vehicles. The only collisions that can occur are between Metro cars.

For these reasons Metro passenger cars do not have to meet the national standards for cars with an 800,000-lb (3,558-kN) compressive loading, but have their own custom design standards. In fact, in the 1970s Metro's cars were built with only a 200,000-lb (890-kN) compressive strength.

Because of this large difference in strength, collisions between Metro trains are not directly comparable to collisions involving commuter or intercity trains. For that reason we are normally not concerned with subway train wrecks. The Washington Metro had three fatal collisions involving overriding and telescoping cars, however, and difficulties resolving the problem.

In 1996, a Washington Metro train had trouble stopping. The train, traveling at 22–29 mph (35–47 km/h), struck a stationary train. The last car of the standing train telescoped the first car of the moving train by about 21 feet (6.4 m). Fortunately, the trains were relatively empty and only the driver was killed. The standing car was crushed by about 10 inches (0.25 m). If the collision had happened during rush hour, there might have been 68 sitting passengers and another 100 or so standing. The cars in this 1996 collision were built in the 1980s with a compressive strength of 300,000 lbs (1,334 kN).

In a similar collision in 2004, an empty Washington Metro train ran away out of control downhill backward and struck a standing train at 36 mph (58 km/h). The first car of the stationary train telescoped about 20 feet (6 m) of the last car of the moving train. The collision debris pushed ahead of the collided car and created a 34-foot (10.36-m)-long crush zone inside the telescoped car. Fortunately, the standing train had just let off its passengers. Twenty people were taken to the hospital.

In June 2009, a Washington Metro train (moving in excess of 35 mph [56 km/h]) had its first car overridden and telescoped about 50 feet (15.2 m) when it struck a stationary train from behind, killing 9. The 1996 and 2009 Metro accidents are described in more detail in Chapter 6.

After the 1996 collision, the National Transportation Safety Board (NTSB) recommended that Metro evaluate the structural design of all its cars for the ability to resist telescoping. Metro hired independent consultants to study the problem. The consultants concluded that "it was neither practical nor desirable to add underframe reinforcement and that such modifications possibly could result in more injuries."[6]

Without the benefit of reviewing the consultants' report, it can be stated that stronger is not always better—especially with respect to crashing. Consider protecting your head from a hammer blow, first with a strong block of steel and then with a similar block of bubblegum. Obviously, the steel block is stronger than the bubblegum, yet the bubblegum does a better job of protecting the head. Why? The definition of work and energy offers an explanation. Work equals force times the displacement it acts over. The steel absorbs little of the hammer's energy of motion and passes the impact force straight through to the head. The bubblegum deforms and absorbs more energy. Apparently, the consultants could not decide if reinforcing Metro's passenger cars was making the structure more like the steel block or the bubblegum. And why couldn't they decide?

Consider the Federal Railroad Administration's (FRA) research to increase the crashworthiness of passenger cars. It started with studies in the mid-1990s that led to many years of crash-testing and advanced computer simulation. Finally, a new and better design was developed that was verified by full-scale crash-testing in 2006. The consultants did not have a similar mandate and budget for 10 years of analysis and testing.

Washington Metro's latest short-term plans call for sandwiching the cars most susceptible to telescoping in the middle of trains made up with more crashworthy cars. This will occur until the Metro can replace its fleet with cars designed to crush on purpose to protect the passengers.

PASSENGER CAR CRUSHING

Passenger car crushing is a good thing, as long as no one inside is being crushed. A controlled crush is the best way to dissipate the crash energy.

The traditional design of passenger cars has focused on making the cars as strong as possible within weight and operating constraints. The American result has been the 800,000-lb (3,558-kN) compressive strength requirement.

An alternate approach designs the ends of the cars to crush on purpose, absorb the crash energy, and protect the passenger space. The crush zones deform absorbing energy just as the bubblegum absorbed the hammer blow. The concept, around since at least the 1850s, has been resurrected and redeveloped with the aid of modern computer simulation and up-to-date crash test technology.[7] Today this method is known as crash energy management (CEM). The British and French were the first to advance the science and deployed CEM passenger cars in the 1990s.

The American version of CEM retains the 800,000-lb (3,558-kN) strength in the middle of the car. The ends of each car contain crush zones designed to crush at a lower force. The unoccupied ends progressively crush to distribute the crash energy along the length of the train. In 1999, a CEM design was first required by the FRA for all U.S. passenger equipment operating above 125 mph (201 km/h), including Amtrak's new European-designed 150-mph (241-km/h) Acela.[8]

The first step of any crash energy management design process is to define a crash scenario. For Amtrak's Acela the crash scenario is a head-on collision at 30 mph (48 km/h) into an identical stationary train. The engineering analysis predicts zero serious injuries for this 30 mph collision, considerably better than existing cab car designs. The outcomes for higher-speed collisions are also expected to be much safer than traditional designs.

This may not sound like much for a train traveling at 150 mph (241 km/h), but consider the following. It is impossible to safely dissipate the kinetic energy of a high-speed train. Many collisions do, in fact, occur at lower speeds when entering or leaving stations. Also, consider the many fatalities described in the beginning of this chapter from low-speed collisions. The European requirement for their more sophisticated trains traveling about 155 mph (250 km/h) is survivability for a similar collision at a speed of only 22.5 mph (36 km/h).

Other crash requirements specified for the Acela (and all other trains traveling above 125 mph [210 km/h]) are as follows. The front end of the power car is designed to crush 40 inches (1 m). The deceleration in any of Acela's passenger cars must not exceed 8 Gs, or eight times the acceleration of normal gravity. Also, the passenger's impact with the seat in front must not exceed 25 mph (40 km/h). No passengers are allowed in Acela's lead car. To protect the crew, the underframe of the cab must withstand a static compressive load of 2.1 million lbs (9,341 kN). At least 13 megajoules (9.6 million foot lbs), or about 25% of the kinetic energy of the 624-ton (566-MT) Acela traveling at 30 mph (48 km/h), must be absorbed by crushing of unoccupied volumes. This includes 5 megajoules (3.7 million foot lbs) in the front of the power car, 3 megajoules (2.2 million foot lbs) in the back end of the power car, and 5 megajoules (3.7 million foot lbs) in the front end of the first passenger car.

TEN YEARS OF CRASH-TESTING AND COMPUTER ANALYSIS sponsored by the FRA set out to design a more crashworthy commuter train. The same sequence of testing described earlier (a single-car crash into a solid wall at 30 mph [48 km/h], a two-car crash, and a cab-led train at 30 mph [48 km/h] crashed into a stopped locomotive) was repeated with the new crash energy management–designed equipment. The results were very encouraging.

The single-car conventional equipment crushed about 6 feet (1.8 m) compared to 3 feet (0.9 m) for the CEM-designed car. The 2-car conventional equipment tested at 26 mph (42 km/h) crushed 6 feet (1.8 m), lifted up 9 inches (0.23 m), and shifted laterally 30 inches (0.76 m). Both trucks near the coupling derailed. The uplift is the early stages of overriding, and the lateral shift is the beginning of lateral buckling that can eventually lead to side-to-side car impacts.

In comparison, pushback couplers designed for the two-car CEM crash test allowed all wheels to remain on the track and the cars to remain in-line when crashed at a higher speed of 29 mph (47 km/h). The pushback couplers are designed to bring the ends of the cars together uniformly with no offset or saw-tooth buckling. Pushback couplers, in common use on lighter transit cars, were redesigned for the heavier commuter cars used in the FRA test crashes. The pushback couplers get out of the way during a collision by pushing back into a pocket filled with energy-absorbing honeycomb crushable material. This allows the car ends to contact uniformly over a bigger area and trigger each car's controlled crush. The car ends' coming together uniformly greatly reduces saw-tooth buckling and the sideway forces that knock the cars off the tracks. Most important, the pushback couplers reduce offsets that lead to overriding and telescoping.

In 2006, the Federal Railroad Administration crash-tested a cab car–led 5-car train designed with crash energy management. The train collided at 30 mph (48 km/h) into a standing locomotive. Fourteen feet

(4.26 m) of crush was distributed across the 5 cars with no loss of passenger or operator space. This compares to an identical conventional equipment crash test in which the cab car overrode the locomotive by 22 feet (6.7 m), crushing the space for 47 passengers. This successful test crash at 30 mph should also be compared to the 3 fatal cab car collisions described in this chapter with approach speeds of 23, 30, and 31 mph (37, 48, and 50 km/h).

COMPUTER SIMULATION CONCLUDED that a cab car–led train of traditional equipment is injury free with a head-on collision with a 13-mph (21-km/h) approach speed. A locomotive-led train of conventional design is safe at 25–30 mph (40–48 km/h). A cab-led train of CEM design has a safe collision speed of 30 mph (48 km/h) with traditional interiors and 32 mph (51 km/h) with modified interiors. A locomotive-led train with CEM passenger cars is considered injury free at 35 mph (56 km/h) and at 40 mph (64 km/h) with the latest safety developments in interiors. Reduced overriding and injuries, compared to traditional equipment, is expected at higher speeds.

IN 2005, after two fatal accidents in 2002 and 2005 (and before its worst accident ever in 2008) Los Angeles Metrolink asked the FRA to use its latest research and help define specifications for new CEM passenger cars. The specifications included a defined crash scenario (head-on with approach speed of 25 mph [40 km/h]), a cab-end crush zone capable of absorbing 3 million foot lbs (4,067 kJ) of crash energy, maximum secondary impact of 25 mph (40 km/h) for the passengers inside, and other performance criteria. The trick is to specify something that someone will be able to successfully design and build. If a collision speed of 60 mph (97 km/h) is required, no company will respond with a proposal—let alone an actual product.

Metrolink took delivery of the nation's first commuter train cars designed with crash energy management in 2010. The new cars had crush zones on each end, collapsible tables, higher seat backs, pushback couplers, and many other new features. The cars were designed and manufactured in Korea. Elsewhere, crash energy management passenger cars are rapidly becoming the norm for new passenger cars. Replacement of all Washington Metro cars with new crash energy management–designed cars is considered the ultimate fix for its problem with overriding.

OTHER COLLISIONS

The most common train collisions occur at road crossings or with something on the tracks. Usually, because the train weighs significantly more than any road vehicle, a collision at a crossing is a nonevent for the train. During the 20-year period beginning in 1986, there were nearly 1,000 railroad collisions with vehicles or track obstructions, resulting in 34 passenger fatalities. During the same period, there were only 48 inline collisions but 27 fatalities.

Trains can also collide when tracks cross or merge. Surprisingly, side collisions are more common than inline collisions. The inline collision is the most violent, the most studied, the most dangerous, the most interesting—and the major purpose of our discussion here.

4

Freight Train Collisions

H ead-on collisions almost always result from the human error of placing two trains on the same track. Many collisions happen on curves, where the line of sight (and the ability to stop) is limited. Usually, the emergency brakes are activated with little effect. Post-crash fire from spilled fuel is always a concern. The crew may decide to jump ahead of the collision.

On May 28, 2002, two freight trains collided head on in dark territory (tracks without signals). This collision occurred on a single track on which trains routinely operated in both directions. On tracks like this one, a remote dispatcher coordinates train meets by issuing "track warrants." Collisions are avoided by the dispatcher directing one of the trains off of the main track onto a siding.

A coal train departed Amarillo, Texas, at 7:40 a.m., heading east. The 6,380-foot (1,944-m)-long train had 116 cars, weighed 15,483 tons (15,731 MT), and was powered by 3 locomotives—2 at the head end and 1 in the rear.

The coal train's first track warrant, issued at 7:47 a.m., granted the train permission to enter the main track. A second warrant required the coal train to wait on the main track near a siding to meet an oncoming freight train headed west. When a heavier train meets a lighter train head on, common practice dictates that the heavy train stops on the main track. The lighter train passes the heavier train by entering a siding. The meeting between these two trains took five minutes, starting at 8:30 a.m. The track warrant also instructed the coal train to continue eastward toward Ashtola Siding and to stop for another head-on meet with an intermodal train.

The westbound intermodal train (7,033 feet [2,144 m] long, 5,545 tons [5,634 MT], with 2 locomotives) was granted authority by the dispatcher to enter the same track as the coal train at 8:26 a.m. The intermodal train's track warrant directed it onto Ashtola Siding to move past the oncoming coal train at approximately 8:50 a.m.

The eastbound coal train received a third track warrant giving new instructions at 8:43 a.m. This warrant authorized the coal train to continue east to its next destination and also repeated the previous instructions to stop at Ashtola Siding and wait for the oncoming intermodal train to pass. At 8:44 a.m. the coal train engineer started a 10-minute cell phone call and missed the stop at Ashtola Siding.

The collision occurred at 8:54 a.m. 7.8 miles (12.5 km) past the Ashtola Siding stopping point. The investigators concluded that the coal train was traveling at 49 mph (79 km/h) when it hit the emergency brakes just 1,093 feet (333 m) before impact; the intermodal train braked at 42 mph (68 km/h) and 1,064 feet (324 m) before impact. The emergency brakes had little effect in so short a distance.

Line-of-sight testing after the collision indicated that the intermodal train had just passed a curve and could only see the coal train about 1,200 feet (366 m) before the collision. The coal train was operating on straight track for many miles.

Both leading locomotives were destroyed. The coal train's locomotive cab section separated from its frame and crushed into an unrecognizable shape. The first 23 coal cars were derailed and destroyed, as were the 2 locomotives and first 3 cars of the intermodal train.

All four crew members jumped before the collision. The coal train engineer and conductor received critical injuries, and the intermodal engineer was killed by derailing cars. The intermodal conductor received minor injuries. The engineer and the conductor of the coal train were both faulted for failing to stop as instructed by the second and third track warrants.

HEAD-ON COLLISIONS

Head-on collisions are the most violent, with the highest crash forces per Newton's Law, $f = m \times a$.

Acceleration is how fast the speed changes. Acceleration can be positive (speed increasing) or negative (speed decreasing). Negative acceleration is also called deceleration.

An object falling due to gravity is the most familiar and observable example of acceleration. The proverbial apple, when dropped from rest, is speeding 32 ft/sec (9.8 m/s) after 1 second, 64 ft/sec (19.6 m/s) after 2 seconds, 96 ft/sec (29.4 m/s) after 3 seconds, and so forth. Acceleration causes the apple to increase its speed by 32 ft/sec (9.8 m/s) every single second and is said to be 32/ft/sec/sec, or 32 ft/sec² (9.8 m/s²).

The speed of any falling object equals 32 ft/sec² times the number of seconds the object has been falling. If the apple has been falling for 10 seconds, its speed (ignoring air resistance) is 32 ft/sec² × 10 seconds = 320 ft/sec.

The average acceleration can also be calculated by dividing the change in speed by the period of time the change occurs in. If the falling apple starts at zero speed and 2 seconds later is falling at 64 ft/sec, the acceleration of the apple is 64 ft/sec divided by 2 seconds = 32 ft/sec².

If a train is traveling at 30 mph (44 ft/sec [13.4 m/s]) and crashes to a stop in 0.2 second, the average rate of change of velocity equals 44 ft/sec divided by 0.2 second, or 220 ft/sec² (67 m/s²). If a locomotive weighs 400,000 lbs (181,437 kg), the average crash force to decelerate the locomotive to a stop in 0.2 second equals the locomotive's mass times the deceleration:

$$\frac{400{,}000 \text{ lb}}{32 \text{ ft/sec}^2} \times \frac{220 \text{ ft}}{\text{sec}^2} = 2{,}750{,}000 \text{ lb of crash force } (12{,}232 \text{ N})$$

Note how the weight of the locomotive must be converted to mass.

The actual crash pulse is far more complicated and changes greatly as individual components of both locomotives crush, fracture, and fold. Nevertheless, this simple calculation is a valid approximation for the average crash force and can be used to illustrate many other principles.

Weight versus Mass

The mass of an object is related to its weight by Newton's Law $f = m \times a$. The force of gravity is the object's weight. Newton's Law becomes weight = mass $\times$ gravitational acceleration. The mass of an object in English units is the object's weight divided by the acceleration of gravity. The mass of a 100-lb weight is 100 lbs divided by 32 ft/sec^2 = 3.125 lb sec^2/ft. The awkward units are required to balance the units on both sides of the equation $f = m \times a$. In the English system, mass is defined in terms of force and acceleration. In the metric system, the newton (force) is defined in terms of mass and acceleration. The weight of 1 kilogram accelerated 9.8 m/s^2 = 1 kg $\times$ 9.8 m/s^2 = 9.8 newtons. The units for mass in the English system are lb sec^2/ft. The units for force in the metric system are kg m/s^2. The two systems are fundamentally different in that they differ in which of the three terms in $f = m \times a$ are defined in terms of the other two.

The faster a locomotive crashes to a stop, the higher the crash forces. If the locomotive crashes to a stop in half the time, or 0.1 second, the average deceleration and the average crash force will both double.

Consider a very rigid wall. (A rigid wall used for crash-testing is described in note 5, Chapter 3.) A locomotive crashing into the rigid wall will stop abruptly. How quickly is difficult to say, but certainly more quickly than if the locomotive crashed into an automobile. Most likely, the auto will barely slow the locomotive. There will, of course, be crash forces when the locomotive hits the car, but they are very small forces compared to those involved in a collision with the rigid wall.

Common sense tells us that the destructive forces of a locomotive crashing into a completely rigid wall will be far greater than those of a locomotive striking an automobile. Our common sense can now be supplemented with the physics of crashing. The rigid wall causes the locomotive to stop faster with greater deceleration compared to a locomotive striking an auto. The greater deceleration requires a higher crash force per $f = m \times a$.

Compared to the crash forces involved with a collision with a small car, the crash forces increase if a larger car is hit and are larger still if a truck is struck. The collision forces become even larger if the freight locomotive hits a 100,000-lb (45,359-kg) passenger train car, still greater if it collides with a 270,000-lb (122,470-kg) commuter train locomotive, and the highest if the freight locomotive collides with another freight locomotive weighing more than 400,000 lbs (181,437 kg). The larger, stronger, and heavier the struck object is, the greater the crash forces. Therefore, a locomotive colliding head on with another locomotive is the most violent collision possible.

One might think colliding with a 100-car, 15,000-ton coal train would be the worst possible scenario. It turns out that the crush force of a 400,000-lb (181,437-kg) locomotive (mostly steel) is far greater than the crush force of a freight car. The trailing cars collide into each other in serial fashion with lower crash forces than the heavier, more solid locomotives. Therefore, collision severity is mostly controlled by the number of locomotives colliding. One locomotive colliding with two locomotives will create greater crash force than one locomotive colliding with another locomotive. Recall the 270,000-lb Metrolink locomotive that was struck by two 429,000-lb locomotives in the collision described in Chapter 2.

Newton's Third Law

For every force there is an equal and opposite force.

When two 439,000-lb (199,127-kg) freight train locomotives strike a 270,000-lb (122,470-kg) commuter train locomotive, both trains collide with the same crash forces. The same crash force applied to a 270,000-lb structure is expected to provide more deceleration than when applied to a 429,000-lb structure—exactly what happened in the 2008 California accident. It's like watching a football player's head jerk backward (to judge how hard the hit was) after confronting a bigger linebacker. The 270,000-lb Metrolink locomotive also had less steel structure to withstand the crash force and therefore crushed more than the freight train locomotive.

THE DYNAMICS OF HEAD-ON COLLISIONS

A football bounces funny and so do train cars during a collision. During a head-on collision, the kinetic energy of motion has to go somewhere. The two locomotives can crush or compress, or one train can shove the other backward. The locomotives can also lift vertically or tilt sideways, as shown in Figure 4.1.

The locomotives can also override each other. A head-on collision

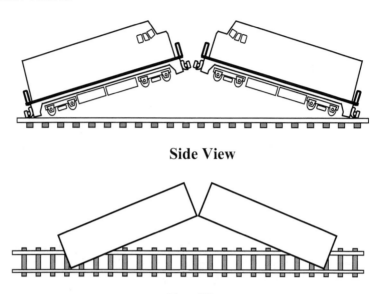

Side View

Top View

FIGURE 4.1. Vertical and lateral buckling of two locomotives during a head-on collision.

involving two identical locomotives traveling at the same speed is a mirror image of itself. In other words, the right half is identical to the left half (Figures 4.2 and 4.3). If the symmetry is maintained during the collision, the two trains will crush identically. Minute misalignment caused by difference in wheel wear, manufacturing tolerances, and/or unequal braking, however, usually prevent symmetry. For example, if the back or front brakes grab more, the front of each locomotive will rotate slightly up or down on its spring-mounted suspensions.

Also, the parts on each train do not fracture, crush, or collapse identically. Even identically made parts will not have identical strength. Something breaks first, disrupting the symmetry. As parts begin to break, one locomotive may ramp up or override the other, as shown in Figure 4.4. Override occurs when the underframe of one locomotive rides on top of the underframe of the other locomotive.

The underframe is the backbone of the locomotive and its strongest part. The underframe must be strong enough to support the engine, pull long trains, and survive repeated coupling forces. Given a typical 74-foot

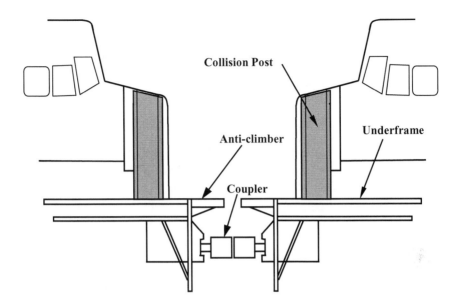

FIGURE 4.2. Structural elements of two locomotives in a head-on collision.

FIGURE 4.3. Top view of the front end of a locomotive, highlighting the collision posts protecting the cab.

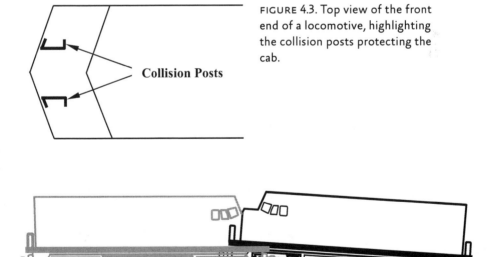

FIGURE 4.4. Overriding locomotives.

(22.5-m)-long, 15-foot (4.5-m)-high, 10-foot (3-m)-wide, 415,000-lb (188,240-kg) locomotive design, the underframe is 87,000 lbs (39,462 kg). Remove the fuel tanks, power plant, trucks, and other miscellaneous equipment, and the underframe makes up about 70% of the remaining steel.

Once override begins, the stronger underframe of the overriding locomotive will crush through the cab of the locomotive being overridden. The crew in the overridden locomotive is in grave danger. The rule-of-thumb requirement for survival is less than 5 feet (1.5 m) of cab crush. Overriding is a particularly insidious event because the crew compartment is crushed with much lower crash force than will occur with uniform crush. The lower crushing force occurs because the stronger underframe is slicing through the weaker parts of the other locomotive.

In 1990, collision posts (see Figures 4.2 and 4.3) on locomotives were required for the first time to have specific strengths. At that time each collision post was required to resist at least 500,000 lbs (2,224 N) at the underframe and 200,000 lbs (889 N) 30 inches (76 cm) above the underframe.

In 1987, 2 locomotives traveling at 9 mph (14.5 km/h) collided head on with 5 locomotives moving at 21 mph (33.8 km/h) for a closing speed of 30 mph (48 km/h). Although the locomotives had collision posts, they did not meet the 1990 standards. One locomotive was overridden with about 7–8 feet (2–2.4 m) of crush, killing 1 crew member.

In 1994, a single locomotive moving at 25 mph (40 km/h) collided head on with 3 locomotives traveling at 18 mph (29 km/h) for a closing speed of 43 mph (69 km/h). The first of three locomotives was overridden. This locomotive had stronger collision posts that met the 1990 standards; the crush was only about 2 feet (0.6 m). There were only minor injuries.

In the 2008 Metrolink head-on collision (approach speed of 84 mph [135 km/h]; see Chapter 2) between two 429,000-lb (199,127-kg) freight train locomotives and a 270,000-lb (122,470-kg) commuter locomotive, all locomotives had 500,000-lb (2,224-N) collision posts. There was no loss of livable space in the freight train's cab compartment. The cab of the commuter train was completely crushed.

The 1990 design standards required an anti-climber (see Figure 4.2)

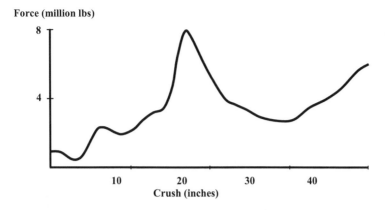

FIGURE 4.5. Computer simulation of the crushing strength of a locomotive with 500,000-lb (226,796-kg) strength collision posts without overriding. *Adapted from DOT/FRA/ORD-03/05.*

to resist a vertical load of 200,000 lbs (90,718 N). It was hoped that the anti-climber would resist locomotive overriding, but experience shows the anti-climber is usually crushed during a head-on collision with another locomotive. The anti-climber is expected to stop smaller road vehicles (and hopefully other freight cars) from ramping up and crashing through the locomotive windshield.

Figure 4.5 shows a computer simulation of a locomotive being crushed at the underframe (i.e., no override). The irregular shape occurs as the metal folds, buckles, and/or fractures.

Computer simulations in the mid-1990s concluded that 1.6 million foot lbs (2,169 kJ) of energy were required to lift one end of the locomotive 10 feet (3 m), to crush the front end about 4 feet (1.2 m) above the underframe, or to crush the front end about 1 foot (0.3 m) at the underframe. This same energy will accelerate the locomotive to about 11 mph (17.7 km/h).

In another computer simulation with 500,000-lb (2,224-N) collision posts, a locomotive struck another standing locomotive at 50 mph (80 km/h) head on and generated a crushing force of 5 million lbs (22,241 N). This simulation assumed a small offset that created a few inches of override. The computer model mostly showed crushing of the underframes of both locomotives.

In 2009, after extensive studies, the strength of each collision post was increased to 750,000 lbs (3,336 N) at the underframe and 500,000 lbs (2,224 N) 30 inches (0.76 m) up. The anti-climbers were also strengthened.

One might falsely conclude that the collision posts can never be too strong. When collision posts are too strong, however, the failure point is moved into the underframe. If the collision posts survive and the underframe folds or fractures, the collision posts connected to a bent underframe could form a ramp that aims the overriding locomotive directly into the other locomotive's structurally weaker windshield. The underframe could be reinforced, but doing so will only create new and different problems. It would be like reinforcing your car with 4,000 lbs (1,814 kg) of armor to provide added protection in a collision. That will work, but at the expense of all cars being struck. (This is basically what happens when large cars collide with small cars.)

In one accident the collision posts survived but failed to protect the cab compartment. The train crew of a passenger train in Canada noticed an incorrectly set track switch ahead. The hand-operated switch, a few feet high with 2 red markers (8 inches [20.3 cm] in diameter), was barely noticeable. This track switch was rarely used and was not designed to be seen at 80 mph (129 km/h). (The markers were visible to investigators actively looking from a slow-moving locomotive at 1,400 feet [427 m]. The train crew members were not at fault but were, in fact, heroes for managing to warn an approaching passenger train.)

Traveling at 80 mph (129 km/h), the passenger train began emergency braking just 600 feet (183 m) from the switch. The incorrect switch forced the slowing train onto a crossover (speed limit 15 mph [24 km/h]) at 74 mph (119 km/h). The passenger train flew off the tracks at 68 mph (109 km/h) and collided with standing freight cars on an adjacent track.

The locomotive struck a loaded hopper car. The severe impact destroyed the front quarter of the locomotive above the underframe, crushed the right side for about half its length, and killed two crew inside (the only fatalities). Seventy-seven of the 186 remaining passengers and crew were injured.

The right collision post was bent but remained attached. The connection between the right post and the underframe was partially fractured. The left collision post was not visibly deformed but had begun to fracture. The investigators concluded that there was no practical collision post design that would have protected the crew from this severe impact.

NEW REQUIREMENTS NOW PROVIDE ADDITIONAL PROTECTION for the crew from collisions with track obstructions. Specifically, the collision posts have to withstand a 30-mph (48-km/h) impact with a 65,000-lb (29,483-kg) cylinder centered 30 inches (0.76 m) above the underframe. To pass the test there must be no more than 24 inches (0.61 m) of crush. The locomotive must also impact an offset 65,000-lb block with no more than 60 inches (1.5 m) of crush. For both impacts intrusions into the cab must be 12 inches (0.3 m) or less. Today these crash requirements are evaluated with computer simulation.

It would be helpful if all crash investigations would clearly record any override, amount of crush, strength of the collision posts, and whether or not the crew was in a defensive position. (The best location is on the floor in the back of the cab away from the crush zone and the weaker windshield.) Unfortunately, this information is usually not available. Not all collisions are investigated, and those that are focus on the cause of the collision and not on the crashworthiness of the locomotives.

A very indirect measure of crash severity is the number of injuries and fatalities. There are many additional threats to the train crew, however, besides being crushed. The crew must survive the environment after the collision. There may be a fuel leak and serious fire or hazardous fumes. The crew can even be electrocuted or drowned.

Survival also depends on secondary impact. After the locomotive crashes to a stop, the people inside continue moving forward per Newton's First Law (a body in motion tends to stay in motion), creating a secondary impact. Exactly how any secondary impact might occur depends on the location of the crew during the collision. Was the crew sitting at their work console, on the floor, or wandering around and thinking about jumping? Even if the crash investigators focus on survival issues, the collision could have thrown the crew out of the cab, making it impossible to assess their pre-crash location.

If the crew applied the emergency brake, this gives some hint about where they were. Applying the emergency brake indicates awareness of the impending collision and that the crew probably had time to get on the floor. Not applying the emergency brake may imply the crew was surprised and that the collision occurred while they were sitting in their seats. Although this is an imperfect indicator, one might expect higher survival rates when the emergency brakes were thrown and the train crew was on the floor. Because the event recorder records if the emergency brakes were triggered, this information is usually available. Occasionally, the investigators describe the position of the crew.

EXAMPLE COLLISIONS

In this section and the ones that follow, we look at example collisions and their survivability outcomes without sorting out the details of each accident. With few exceptions, the major expected trends hold: low-speed survival is very good, high-speed survival less so, with a mixed bag in between.

A common collision scenario involves one train that is stopped or standing. Examples of head-on collisions with a standing train and the emergency brakes being applied without any injuries occurred at speeds of 22, 25, and 38 mph (35, 40, and 61 km/h). In another accident a train traveling at 42 mph (67 km/h) collided head on (without emergency brakes) with a standing train and all train crew survived the collision. One crew member later died from leaked chlorine gas from a derailed tank car.

Other accidents with both trains moving and colliding head on had less positive outcomes. All train crew died at colliding speeds of 51 and 23 mph (82 and 37 km/h), 23 and 45 mph (37 and 72 km/h), 60 and 10 mph (97 and 16 km/h), and 54 and 20 mph (87 and 32 km/h). These collisions represent combined approach speeds of 74, 68, 70, and 74 mph (119, 109, 113, and 119 km/h).

At intermediate approach speeds, 2 of 4 train crew died in a head-on collision occurring at 30 and 13 mph (48 and 21 km/h); the emergency brakes were not applied. One of 4 died with 2 trains colliding at 30 and 10 mph (48 and 16 km/h). All crewmen survived a head-on collision at 22 and 30 mph (35 and 48 km/h).

THE CREW JUMPS

If a freight train crew is aware of an impending collision, they face a very difficult decision: stay with the train or jump off. Because freight train cars can potentially derail on top of anyone who safely jumps, the decision to jump or stay is not clear.[1]

It is usually not clear (or recorded) if a crew member dies during the jump, as he or she lands, or after he or she is thrown from the train during the collision. The crew member might even be hanging on the railing outside the locomotive—probably the worst place to be—while he or she hesitates about jumping. The decision to jump may also be influenced by what is on the train and how hazardous it is. At the crash recounted at the beginning of this chapter, all crew jumped from two colliding trains.

In one example 2 trains collided head on at 5 and 43 mph (8 and 69 km/h); both trains threw their emergency brakes; and 3 locomotives and 31 cars derailed. The conductor jumped at 43 mph and died. The engineer stayed and lived. The other crew was uninjured.

During a rear-end collision in 1997 with a standing train, the engineer placed the train in emergency at 45 mph (72 km/h). The engineer and the conductor both jumped at 20–25 mph (32–40 km/h); one lived and the other did not.

A more common scenario is one train striking the side of another train at a crossing or where tracks merge. One such accident happened in 1998. Three railroaders jumped ahead of impact at 30 mph (48 km/h); 2 lived and 1 did not.

There is one exception to both crews activating emergency brakes before an impending collision. If one train is already planning switching off the main track and onto a siding, that train may instead try to speed up and get off the tracks ahead of the collision. It is difficult for a long train to quickly get off the main track, but at least a more serious head-on collision is avoided. The side collision still has great potential for serious derailment and the colliding train crew is still in danger, but at least the collision will be a glancing blow.

In 2002, just such a collision occurred. The engineer and conductor of the first train jumped off their train at 47 mph (76 km/h) ahead of the collision and received serious injuries. The second train raced to its turnoff and just barely avoided a head-on collision. The second train's crew avoided all injury when their train was struck in the fourth car behind their locomotives.

In 1996, an eastbound train traveling 36 mph (58 km/h) did not slow down and struck a westbound train 25 cars behind the locomotive at a crossing. Sitting in their seats, both crew members of the eastbound train died. In 2003, another side collision occurred at 49 mph (79 km/h). The train crew did brake at the last second and survived with serious injuries. Other examples of side collision occurred at 36 mph (58 km/h) with both crew killed and at 40 mph (64 km/h) with no injuries.

In 2004, a train crew survived a glancing-blow side collision at 44 mph (71 km/h). An eastbound 123-car BNSF freight train exited the main track onto a siding. Meanwhile, a westbound Union Pacific (UP) freight train passed a signal at 44 mph and obliquely struck the midpoint of the

FIGURE 4.6. A UP train collides with the midpoint of a BNSF train. The UP train derailed 4 locomotives and the first 19 cars. The BNSF train derailed 17 cars. *EPA.*

BNSF train. Four locomotives and the first 19 cars of the striking UP train derailed, as did 17 cars of the BNSF train (Figure 4.6). Aware of an odor, the engineer and the conductor of the UP train evacuated after the collision on foot. Unfortunately, the punctured 16th car of the UP train released 120,000 lbs (54,400 kg) of chlorine gas. The evacuating UP conductor and 2 residents (220 feet [67 m] away) were killed by the chlorine vapors. The danger zone was considered to be at least a 700-foot (213-m) radius. (A major chlorine release is described in Chapter 8.)

REAR-END COLLISIONS

Unless a freight car bounces up and crashes through the cab, as shown in Figure 4.6, rear-end collisions are expected to be less dangerous than a head-on collision because there is less colliding mass.

Again, the most common scenario is the collision of a moving train with a stopped train. Rear-end collisions with stopped trains were recorded with no injuries at speeds of 20, 24, 28, 35, 38, 40, 45, 48, and 53 mph (32, 39, 45, 56, 61, 64, 72, and 85 km/h). The crew of the 53-mph (85-km/h) colliding train did apply the emergency brakes and get into the brace position; 3 locomotives and 33 cars were derailed. The train colliding at 45 mph (72 km/h) derailed its 3 locomotives and 10 cars.

In another example 2 railroaders were killed in a rear-end collision without braking and colliding at just 23 mph (37 km/h). In this accident the left collision post was bent backwards 42 degrees from impact damage 39 inches (1 m) above the underframe. A flat car of the struck train crushed the crew compartment of the striking locomotive. One end of the flat car ended up on top of the heavily damaged locomotive.

A train traveling at 56 mph (90 km/h) in foggy conditions rear-ended a train traveling in the same direction at 8 mph (13 km/h) for an approach speed of $56 - 8 = 48$ mph (77 km/h). There were 2 fatalities on the 56-mph colliding train. The collision derailed 3 locomotives and 13 cars. A third train collided with the derailed cars, derailing another 18 cars. There were no injuries on the other two trains.

One person died in a rear-end collision at just 30 mph (48 km/h). The last car of the train that was struck was a flat car that catapulted into the cab, as shown in Figure 4.7. This type of collision is particularly dangerous because the windshield is a natural weak point. In a crash test

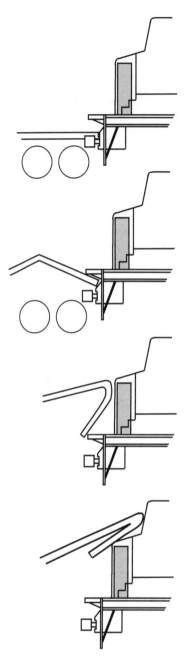

FIGURE 4.7. The underframe of a flat car overriding a locomotive during a rear-end collision. Unlike passenger cars, freight cars easily lift off their trucks.

designed to study this mechanism a locomotive collided at 40 mph (64 km/h) into a standing train with 30 loaded hopper cars and a flat car on the end. In this case the flat car did not catapult into the windshield; instead, the locomotive overrode the flat car. A locomotive overriding a flat car should be a survivable event.

In another crash test a locomotive trailing 3 loaded hopper cars collided with a stationary train of 35 loaded hopper cars at 32 mph (51 km/h). The locomotive overrode the last hopper car with zero crush damage to the crew compartment.

BEFORE MODERN COLLISION POSTS

Reviewing collisions from the 1960s through the 1980s clearly demonstrates the benefit of locomotives with modern collision posts.

In 1981, a 247,000-lb (112,000-kg) freight train locomotive traveling at 12 mph (19 km/h) collided head on with an 85,000-lb (38,555-kg) commuter cab car. The commuter car was being pushed at 19 mph (30 km/h) by a tail-end power car. Surprisingly, the lighter commuter car did not get crushed but instead bounced up and overrode the locomotive, killing three inside. Apparently, before sufficient protection was provided with stronger collision posts, lighter equipment often overrode the heavier locomotives.

There are certainly examples of wooden cabooses being crushed and the occupants killed, but often a steel caboose will bounce up and crash into the locomotive cab. (The caboose has been replaced by modern sensors since the 1980s. See Chapter 8.) Such accidents were found at collision speeds of 9, 11, and 15 mph (14, 18, and 24 km/h), each time killing one inside the locomotive. These 3 accidents happened in 1966 (derailing 5 cars and 3 locomotives), 1969 (derailing 3 cars and a locomotive), and 1970 (derailing 2 cars and a locomotive).

Unbelievably, in 1970, two were killed inside the cab when a freight train and its caboose backed into a standing locomotive at 3–6 mph (5–10 km/h)! The jolt was so gentle the locomotive was not derailed and the caboose was barely damaged. The crew braced themselves, believing there would only be a hard bump. By all appearances, that's all that happened except the caboose bounced up and crashed into the crew compartment.

Another 1970 crash report describes the structural reinforcement of the caboose being almost 5 times greater than that of the locomotive. The locomotive had only 6 steel box beams protecting the cab compartment. Each beam box section was 2 inches × 2 inches (5 × 5 cm) and only 0.125 inch (0.31 cm) thick. Depending on the steel's strength, modern collision posts appear to be perhaps 4 times stronger. Also, anti-climbers did not exist back then.

5

Avoiding Collisions

On October 30, 1972, Train 416, a Chicago commuter train, accidentally overshot the 27th Street station platform by 600 feet (183 m). In defiance of all common sense, the train backed up. At 7:38 a.m. Train 416 was struck from behind by Train 720, killing 45. The structural damage in this accident was previously described (see "Telescoping" in Chapter 3). The signal system, designed to prevent a collision, is described here.

The operating rules for this train actually allow backing up, provided the train is protected by a flagman. The flagman must be far behind the train with a clear view of any approaching trains. A flagman is an old-fashioned concept that has gone out of practice with the use of automatic signals. This train crew had never used a flagman and had forgotten the rule. The conductor leaned out an open door to look back while talking to the engineer through the train's intercom. A curve in the track blocked his view. Just before the collision the conductor shouted at the engineer to stop and jumped off the train.

The signal system, installed between 1926 and 1929, had three colors: green—"all clear," yellow—"slow to 30 mph (48 km/h) and prepare to stop at the next signal," and red—"stop." Train 416 stopped past the signal near the platform (Figure 5.1). Because the train passed the signal (labeled Red in Figure 5.1) while moving forward, the previous signal

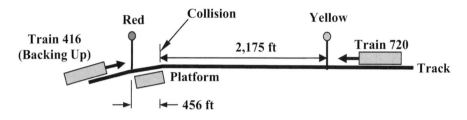

FIGURE 5.1. Train 416 overshoots a red signal by 400 feet (122 m) and backs up on the track. A curve in the track interrupts the line of sight.

(labeled Yellow in Figure 5.1) was reset to yellow. The yellow signal allowed Train 720 to enter the next block of track at 30 mph while preparing to stop at the next signal. When Train 416 reversed and drove past the signal near the platform again, the distant yellow light automatically changed to red. Unfortunately, it was too late for Train 720 to see the signal change and stop.

Train 720 was also speeding and only 2 to 3 minutes behind Train 416. (The trains normally run this close together during rush hour.) Train 720 was an express train and did not normally stop at the 27th Street Station. The normal speed at this location is 65 mph (105 km/h), but slows to 30 mph (48 km/h) when the signal light is yellow.

The engineer of Train 720 testified that he saw Train 416, on the curve, just 4 to 6 car lengths ahead (about 300–400 feet [91–122 m]). The engineer of Train 720 applied the emergency brakes, threw the train into reverse, blew the horn, and left his position to warn the passengers in his car. Both trains were multiple-unit (MU) trains without locomotives. (Each car of both trains was self-propelled with electric motors.)

Train 720 struck Train 416 at an estimated 44 to 50 mph (71–80 km/h). The first car of Train 720 was crushed about 10 feet (3 m) and telescoped about 35 feet (11 m) of the last car of Train 416. Most of the fatalities occurred in the telescoped car.

If Train 416 had not overshot the platform or had chosen instead to continue to the next station or had backed up with proper protection, if there had not been a curve in the track restricting the conductor's view, if Train 720 had not been speeding or operating so closely behind Train 416 or otherwise had a chance to see the yellow signal light change to red, if the two trains had radio contact with a dispatcher, or maybe even if the back of Train 416 had not been painted black with very small marker lights, the collision most likely would not have happened. These were not unlucky coincidences, however, but are instead examples of systemic failure and very sloppy and unsafe operating procedures.

Line-of-sight testing after the fact determined that Train 720 should have been able to see the distant red stop light at more than 1,700 feet (518 m) and the nearly stopped train at more than 1,800 feet (548 m)— more than enough distance to safely stop a 30-mph (48-km/h) train. The investigators concluded that colliding Train 720 was traveling too fast to safely stop.

THREE TYPES OF TRAIN COLLISIONS MUST BE PREVENTED: head-on, rear-end, and side collisions at crossovers (where trains move from one parallel track to another) and crossings (where two tracks intersect).

The locomotive engineer must plan ahead to stop. The fundamental problem is managing the kinetic energy of a long, heavy freight train or a very fast passenger train. Because of curves and/or adverse weather, the stopping distance can easily exceed the driver's line of sight. Most collisions are due to human error and can be easily prevented by following correct procedures.

WHEN RAILROADING FIRST BEGAN, it was easy to separate trains with long periods of time between departures. Even trains moving in opposing directions on the same track could be safely separated. The control system was a strictly followed timetable. As traffic increased, rules were developed to prevent potential collisions. Meetings became prearranged. An "inferior" train cleared the main track on a siding and waited for the "superior" train with higher priority to pass. Stopped trains were protected from a rear-end collision by flagmen. Problems still occurred when new trains were added to the schedule or lengthy delays occurred.

For a system totally dependent on time, surprisingly, nobody could agree what time it was. Every town used local sun time. There were 38 local times in Wisconsin and 27 in Illinois. The Pittsburgh train station used 6 different clocks to track the arrival and departure of trains. In 1883, railroads established Standard Time, and many localities followed. Congress finally agreed to codify standardized time in 1918.

The situation improved somewhat with the introduction of telegraphs. The timetable was still sacred, but could now be modified with written train orders. A dispatcher would issue train orders sent by a telegraph to an operator in the field, who would deliver the orders to the train crew at the next station.

Today train dispatchers issue orders directly to the train crew by radio. The most common version of this system today is known as Track Warrant Control, or TWC, commonly used on low-traffic routes. Under TWC there is no timetable. Track warrants are issued for each section of the trip.[1] The train crew repeats the instructions and writes them on a Track Warrant Form. An accident involving track warrants is described at the beginning of Chapter 4.

BLOCKS

On busy routes two tracks replaced single track. Trains would operate in only one direction on each track, eliminating head-on collisions. Rear-end collisions were still possible.

A manual block system was first developed in England in the 1850s to separate trains traveling in the same direction. An operator controlled the entrance to each block with signals. A following train could enter the same block only after verifying (by telegraph) that the first train had entered the next block.

In 1907, the New York Central operated 226 block houses along the 395 miles (636 km) between New York City and Buffalo. As each train passed the block house, the operator would change the signals to prevent all following trains from entering the same block. The block operator also telegraphed the central dispatcher the position of each passing train.

One dispatcher kept track of perhaps 30 trains and sorted out any operational changes. If a freight train got off the main track at the same time and place every day to allow a faster passenger train to pass, the dispatcher had to adjust for any changes that altered the passing location. The dispatcher telegraphed the new location to the appropriate control tower. Signals and track switches were modified to guide the freight train off the main track at the new location.

In 1907, nearly 50,000 workers were required to operate all the trains to and from (and all points in between) New York City and Buffalo.[2] This included signalmen, telegraph operators, dispatchers, porters, conductors, engineers, firemen, brakemen, switch tenders, and yardmen. Not included were track repair crews, mechanics, and clerks.

A block length is determined by the safe stopping distance of the hardest train to stop, that is, the train with the most kinetic energy of motion. The safe stopping distance depends on the length, weight, and speed of the train; gradient of any hills; and line of sight for signals. This distance can vary from a few hundred feet to many miles.

AUTOMATIC BLOCK SIGNALS

Invented in America in 1872, the automatic block circuit changed railroading forever. In a track circuit a section of track is electrically isolated from adjacent track. A battery is connected at one end to an electromagnetic relay on the other, as shown in Figure 5.2.

A relay is an electromagnetically operated electric switch. The relay turns signal lights on and off. When the electromagnetic coil inside the relay box is energized by the track circuit, the armature pivots and completes the circuit for the green light (Figure 5.3).

If a train enters the block and shorts the circuit, the current flows through the train's steel wheels and axles instead of through the rails; the electromagnetic coil loses its magnetism. Gravity drops the armature and the red light or stop circuit is activated. The presence of the train sets the signal—not an operator.

The Automatic Block System (ABS) protects against rear-end collisions by signaling that the next block is occupied by a train. The automatic block signal circuit is considered "fail safe." If the signal is de-energized by

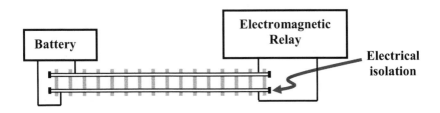

FIGURE 5.2. A battery is connected to both rails. The current from the battery powers an electromagnetic relay that controls additional signal circuits shown in Figure 5.3.

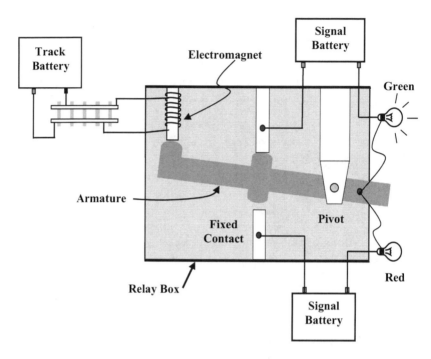

FIGURE 5.3. A relay box shown in the "all clear" position (green light on). The electromagnetic relay attracts the left end of the armature up and turns on the green light.

a broken wire (or dead battery), the signal defaults to its safest setting—"stop."[3] The automatic block signal can also detect a completely broken rail, but not a partially broken rail that might break under the next passing train.

While the block system circuit worked quite well, the actual signal devices were not reliable until small electric motors were perfected around 1900 and the venerated pivot arm semaphore came into wide use. Light bulbs had to wait until the bulbs were bright enough to be seen in the daytime.[4]

The track circuit described is called a two-aspect system—the signal can only describe "all clear" or "stop." A train entering an empty block will receive an "all clear" signal, even if a proceeding train stops just inside the next block. In this case a following train will see the stop signal and stopped train at the same time. The engineer has to anticipate the next signal being a "stop" and slow down or risk collision.

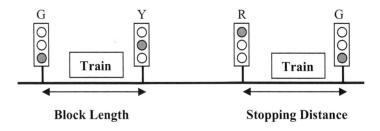

FIGURE 5.4. Three-aspect system. The block circuit keeps track of occupancy in the next two blocks. Green (G), Yellow (Y), and Red (R).

The automatic block signals were rewired to acknowledge trains in the next two blocks. This led to the three-aspect block system (Figure 5.4): red, green, and yellow—"stop," "approach," and "prepare to stop" at the next signal. Nevertheless, the two Chicago commuter trains at the beginning of this chapter managed to collide with three-aspect signals.

Sometimes a four-aspect system is used for faster trains. It keeps track of three blocks and provides a fourth signal for "all clear." An Amtrak train, operating with a four-aspect automatic block system, collided into the back of a freight train at 107 mph (172 km/h), described at the beginning of Chapter 6. The freight train defeated the additional warnings by accidentally entering the track immediately in front of the passenger train.

This is by no means a complete explanation of railroad signals. The Canadian Pacific operating rules list combinations of 13 different lighting configurations to indicate 54 different indications or instructions.

In official railroad signal jargon, a signal aspect is the visual appearance of the lights (i.e., two vertical red lights, two horizontal green lights, a flashing yellow light, etc.), while the signal indication is the actual meaning (i.e., "all clear," "stop," or "proceed at medium speed").

To list just a few (and show how we might arrive at a list of 54): stop and wait for a conflicting train to pass (called absolute stop), come to a complete stop and proceed at slow speed (called permissive stop), prepare to stop at the next signal, prepare to stop at the second signal, proceed at medium speed until the train clears all track switches, proceed on diverging route at prescribed speed through turnout, proceed at reduced speed, proceed at restricted speed, proceed prepared to pass the next signal at slow speed. Slow, medium, limited, reduced, caution, and re-stricted speed are six different speed limits, depending on the situation and the railroad company. Caution speed might mean a speed that per-mits stopping within one-half the range of vision—which can vary greatly with the radius of a curve or in foggy conditions.

AUTOMATIC BLOCK SIGNALS substantially reduced rear-end collisions. The block system was also adopted to increase operating efficiency and pack more trains into a given stretch of track.

Today about 40% of U.S. tracks remain "dark territory," or non-signaled, without automatic blocks (or centralized traffic control, dis-cussed later).[5] Dark territory accounts for about 20% of train traffic and a very small percentage of passenger trains. In dark territory speeds are limited to 49 mph (79 km/h) for freight trains and 59 mph (95 km/h) for passenger trains.

INTERLOCKS

An interlock is a machine that automates a sequence of signals and track switches to prevent two trains from occupying the same track. Interlocking is used at crossovers between two parallel tracks and at junctions where tracks intersect.

Interlocking was first invented in densely populated England in the 1850s. At that time multiple track switches and signals at busy junctions were controlled semi-remotely at one location with mechanical linkages. The next step was to interconnect or interlock the sequence of signals and switches. Interlocking required the switches to be correctly set be-fore the signals could be changed. Levers for conflicting routes could not be pulled at the same time.

Initially, all interlocking was accomplished mechanically with levers in a control tower, as shown in Figure 5.5. The levers would directly con-nect to pipes controlling track switches and pipes or wires controlling signals. A 1925 signaling handbook refers to the thermal expansion of pipes up to 1,300 feet (396 m) long! Shown in Figure 5.6 are the pipes for the mechanical interlocking at Hancock Tower in Hancock, West Vir-ginia. The levers and pipes remained in operation until 2007.[6]

Later, air pressure would transmit a signal perhaps thousands of feet.

FIGURE 5.5. Wilson Tower, Chicago. The tower was constructed in 1900 and operated as a mechanical interlock until 1996. *Photo by Jack Boucher, Library of Congress.*

Remotely located pressurized air cylinders would activate switches and signals. For many decades air cylinders were the preferred method for actuating switches and signals. Until small electric motors became reliable, early electric signals controlled electromagnets that activated remote air cylinders.

SIGNALING REVISITED

There are two main types of signaling: block signaling, to separate trains going in the same direction, and interlocking signaling, to avoid collisions when trains potentially share the same tracks. The two types of signaling require two different types of stop signals—absolute stop and permissive stop.

Permissive stops occur with automatic block signals. The train must stop at the red light. After stopping, the train may proceed at a speed that would allow stopping within the field of view, usually 15–20 mph (24–32 km/h) on straight, level ground, but much slower on curves or during

FIGURE 5.6. Mechanical interlocking pipes at Hancock Tower (Hancock, West Virginia) used to operate remote switches. *Photo Credit: Michael Brotzman.*

inclement weather. Trains were allowed to proceed (at reduced speed) because one faulty stop signal could gridlock the system.

An absolute stop occurs at interlockings where two or more trains can be routed onto the same track. At least one train must stop and remain stopped until the signal changes or the dispatcher gives permission to proceed.

On single track used in both directions one common arrangement is to use permissive stops for all trains going with the normal flow and absolute stops for trains traveling in the opposite direction.

WHEN INTERLOCKING DERAILED TRAINS ON PURPOSE

At the end of the nineteenth century and at least until the 1920s, it was common practice to derail a train to prevent a collision. A device known as a derailer would protect busy track junctions by derailing any errant trains that missed a stop signal. The derailer was interlocked with track switches and signals.

AS TRAINS BECAME FASTER AND HEAVIER, it became harder to control a derailed train's momentum. The train could still reach the intersection, even when derailed by the derailer, as illustrated by the following accident.

In 1921, an eastbound Michigan Central train, the Canadian, left Chicago at 5:05 p.m. bound for Detroit. A westbound New York Central train, the Interstate Express, left Buffalo at 8:30 a.m. bound for Chicago.

The two trains arrived at the same intersection in Indiana every night at about the same time. Usually, the Canadian got there first and had the right of way. On the night of the accident, the Interstate Express got to the intersection first. The interlocking operator in the tower set the signals and derails to stop the Canadian and allow the Interstate Express to pass. The Canadian was derailed 310 feet (91 m) from the crossing. Amazingly, the locomotive re-railed itself and dragged all nine derailed cars until the third car stopped abruptly on the tracks in front of the oncoming train. The Interstate Express rammed into the stopped Canadian at 50 mph (80 km/h) at 6:20 p.m. on February 27, 1921, 40 miles (64 km) southeast of Chicago.

An eyewitness in the fourth car of the Canadian relates the following:

The car swayed suddenly and left the tracks. We humped along the tracks for a few feet at full speed. I looked out; there was another train bearing down on us at full speed. I couldn't move. My tongue stuck in the roof of my mouth. I tried to open the door into the day coach to shout to the people inside; my hand refused to function. It was perhaps not more than thirty seconds between the time I saw that train and the moment it hit us. It was ten years to me. The light from the headlight of the approaching train made everything as bright as day where I stood. I thought I was surely going to be killed.

As I stood I could see into the day coach. On the side from which the New York Central train was approaching I could see that other passengers had seen the train. A woman jumped up from her seat. She held a little girl—looked as though she was about 8 years old—to her breast. Men jumped up—one even started for the door where I was. Then suddenly the side of the car buckled in. My last conscious impression was of everything dissolving in front of me. I remember seeing the nose of the engine . . . the lights in the car going out suddenly . . . hearing a scream that I'll never forget. And then came darkness.

I was thrown from the platform nearly fifty feet (15 m). I fell on the ground, the breath knocked out of me. When I sat up the whole middle of our train had gone. It was a little hell there for a few minutes.

Our eyewitness was uninjured except for a few bruises.

An eyewitness in the back of the New York Central train stated: "Out of the clear sky there came a crash that was like a cannon's roar. The car filled instantly with a cloud of dust. Men and women were hurled to the floor. Our train had been going full speed. As soon as it stopped I went through the car and saw that no one was seriously hurt."[7]

When asked what happened, the surviving Michigan Central engineer said: "The yellow warning at the distant block was against me and we slowed down. But the fireman said the home signal (at the crossing) was clear and I thought it was clear, too."[8]

There were 37 fatalities, including the engineer and fireman of the New York Central train. The third car of the Canadian, with 60 to 80 passengers, was totally destroyed. Most of the fatalities occurred in this car.

A DERAILER ALSO FAILED to stop a passenger train 490 feet (149 m) from a New Jersey drawbridge on September 15, 1958. Forty-eight people drowned when 2 locomotives and 3 passenger cars fell into Newark Bay. The speed limit on the bridge was 45 mph (72 km/h). The engineer drove past two signals. The first signal indicated that the train should reduce speed by one-half and prepare to stop at the next signal. The train was going 43 mph (69 km/h) when it ran through the derailers.

Derailers are still used today to prevent cars on a siding from accidentally entering the main track for whatever reason. In this case the derailers are effective because freight cars move slowly on sidings.

THE MECHANICAL INTERLOCKING BED

Interlocking, a crude mechanical computer with its own logic (i.e., if lever 1 and lever 2 are thrown, lever 3 must be locked in place), was accomplished with a mechanical "locking bed." Each lever in a mechanical interlock directly operates a track switch or signal. The levers also operate bars in the locking bed. A locking bed consists of vertical and horizontal locking bars. When an operator pulls a lever, a vertical bar in the locking bed moves. Attached to the vertical bar is a triangular wedge. This wedge contacts a similar wedge attached to the horizontal locking bar with angled contact so that when the vertical bar slides down, the horizontal bar slides across (Figure 5.7). Also attached to the horizontal locking bars are locking "tappets" that slide into a notch on a vertical bar, preventing its movement. The following logic, provided by the locking bed, is required to prevent a collision between the two trains shown in Figure 5.7.

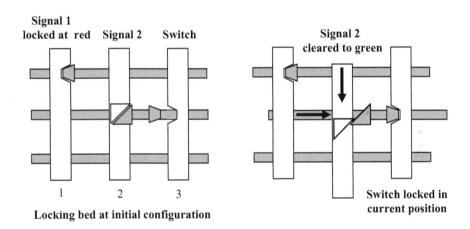

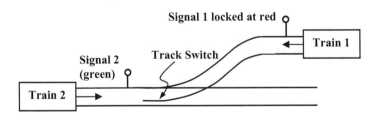

FIGURE 5.7. Simplified example of a mechanical locking bed. Signal 2 has been activated and locks the switch, preventing it from being altered.

1. Signal 1 is locked at red and stops Train 1. Signal 2 is activated by the operator with a lever and cleared to green. The lever also slides vertical bar 2 in the locking down position.

2. The triangular wedge on vertical bar 2 contacts a similar wedge on the horizontal locking bar, forcing it to slide to the right.

3. A locking tappet attached to the horizontal locking bar slides into a notch on vertical bar 3, locking the switch into its current position. The through train safely continues. The switch cannot be accidently changed underneath the passing train.

A very complex sequence of interlocking can be obtained by attaching multiple triangular wedges to each vertical bar that activate multiple horizontal locking bars. And each horizontal bar can have multiple tappets that lock multiple signals or switches.

The locking bed for the levers shown in Figure 5.5 is shown in Figure 5.8.

INTERLOCKING LOGIC IS COMPLICATED

Consider the simple track layout shown in Figure 5.9. Signals 1 and 2 are in their default state of "stop" and directed at approaching Trains 1 and 2. The track switch is in its normal position, allowing trains to pass from A to B.

For Train 2 to safely pass through the switch to point B, the following interlocking has to occur. The switch is thrown to allow Train 2 to proceed safely through the branch. A separate circuit verifies that the switch completed its movement successfully. Signal 1 is locked at "stop." Signal 2 can now change and allow Train 2 to pass. The interlocking logic must also guard against the switch accidentally being changed underneath the passing train. Additional logic is required to coordinate distant signals and different combinations of train directions. The logic becomes intensely complicated when tracks merge, cross multiple tracks, move in both directions, and so on, as shown in Figure 5.10.

Surprisingly, even after electric signals were used to activate distant signals and switches, the mechanical interlocking bed was still used for many years to supply the interlocking logic.

When direct mechanical linkages were eliminated and replaced with electric signals, shoulder-high levers were no longer needed to activate

FIGURE 5.8. The locking bed for Wilson Tower; see levers in Figure 5.5.
Photo by Jack Boucher, Library of Congress.

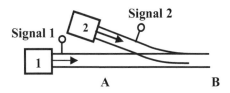

FIGURE 5.9. Simple track switch used to illustrate interlocking logic.

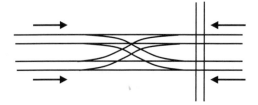

FIGURE 5.10. Example of tracks that require complex interlocking logic.

FIGURE 5.11. Interlocking control panel for Pennsylvania Railroad, Hunter Interlocking Tower, Newark, New Jersey. Behind the levers is the horizontal mechanical locking bed. *Historic American Engineering Record; Library of Congress.*

track switches several hundred feet away. Smaller hand levers were still used to operate the mechanical interlocking bed (Figure 5.11).

GRAND CENTRAL TERMINAL

The most complex electrical interlocking machine with a mechanical locking bed was installed in 1913 in New York's Grand Central Terminal. There were 145 separate tracks at Grand Central: 41 tracks on the upper level, 22 on the lower level, and 62 tracks for train storage. The trains were routed to 67 platforms. Grand Central Terminal was the largest train station in the world at the time.[9] Underneath Park Avenue, the 67 tracks from the 67 platforms merged into 4 tracks that entered and exited the station. Controlling about 1,200 train movements per day (today 1,500), concentrated in the 4 hours of rush hour, were 238 track switches and 570 signals. The upper-lever interlocking machine had 400 levers, and the lower-level machine had 376.

When they were 5.3 miles (8.5 km) from Grand Central, arriving trains were announced by telephone in the control tower. About 30 blocks from the terminal, the train would light a bulb on the control panel. The train director placed each arriving train by calling out lever numbers. An operator (controlling 40 levers) rotated each lever called. The position of each train changed on the control panel, and the director called out another number until the train was correctly placed.

It took about 20 levers (sometimes up to 60) in the correct sequence to park an average train. The levers were mechanically interlocked so they could not be pulled out of sequence. "Imagine a piano whose keys could be so interlocked that each note of each piece had to be played in its proper order."[10] The interlocking machine at Grand Central remained in operation until 1993, when it was finally replaced by modern computer controls.

THE ELECTRIC RELAY

The mechanical locking bed was eventually replaced with electric relays. (Recall that a relay is an electrical on-off switch; see Figure 5.3.) One electrical relay can activate many other electrical relays, each controlling a circuit to activate or deactivate any complex sequence of signals and switches as needed. An elaborate interlocking can involve hundreds, even thousands, of relays.

In 1925, the Interstate Commerce Commission (ICC) reported 7,023 interlocking plants in the United States, with 930 electrical and the rest mechanical. An all-electric interlocking was first used in 1907, but new installations were not widespread until the 1920s. A 2007 survey by the Federal Railroad Administration (FRA) indicated nearly 14,000 interlockings, with only 110 being manually operated. The last all-mechanical

TABLE 5.1
Survey of Interlockings in England, circa 2003

Interlocking Type	Approximate No.	Estimated Average Age
Mechanical	600	75
Relay-based	850	25
Computer-based	250	7

interlocking tower (Ridgely Tower in Illinois), complete with levers and pipes, was deactivated in 2010.

Every electrically controlled track switch and signal required a power circuit to operate the device and a control circuit to process the interlocking logic. The control circuit was wired to perhaps hundreds of relays with complex interconnections to create the control logic. Starting in the 1980s, the control circuits were replaced by a microprocessor. One microprocessor could replace hundreds of relays and greatly reduce installation costs. The microprocessor computer logic was identical to the mechanical interlocking logic from a century earlier.

In a layout begging for computerization, the New York Subway System in 2004 had 856 track miles (1,378 km), 11,646 track circuits, 327,156 relays, and 198 interlockings (70% of the interlockings were all relay, 30% older style electro-mechanical or electro-pneumatic, and 1 solid state). The New York City Transit (NYCT) transports more than 5 million commuters per day and averages 1 "apparent" signal failure every 10 hours. To get a train moving within 10 minutes, NYCT has a crew of nearly 1,000 to maintain signals.

A 2003 U.K. SURVEY gives some insight into the life expectancy of older railroad technology, at least in England. It estimated that there were about 1,700 interlockings in the United Kingdom at the time (Table 5.1).

INTERLOCKING ON THE
NORTHEAST CORRIDOR

The Northeast Corridor (NEC) operates in the most densely populated part of America—a 456-mile (734-km) route connecting Washington, D.C., Baltimore, Philadelphia, New York City, and Boston. The

NEC has been historically and remains the busiest, most technologically advanced, and only example of a high-speed train in America.

In 1975, this busiest train route in America had 124 manned interlockings on its 456 route miles. About 5% of the track miles are interlocking. The trains were handed off from interlocking to interlocking, in what was effectively a manual block system.

As the Northeast Corridor was slowly modernized for high-speed traffic, the number of interlockings changed very little. Instead, they were reconfigured and automated. The mechanical interlocking beds survived for a surprisingly long time on the NEC. In 1975, there were 104; by 1986, there were 25 and just 2 in 2002.

Why are there so many interlockings on the NEC? There are hundreds of bridges on the NEC. Tunnels leave and enter Baltimore and New York City. To reduce costs, four tracks converge to two before crossing a bridge or entering a tunnel, and then diverge back to four upon leaving. Most stations and yards have multiple tracks entering and leaving. Other tracks also need to cross the NEC.

Before modernization, the tracks on the Northeast Corridor were mostly laid out in the mid-nineteenth century and optimized for traffic patterns in the 1940s—the last time the now-bankrupt Pennsylvania Railroad had money for major improvements. In the mid-1970s, there was a hodgepodge of tracks along the NEC, with routes having two, four, or six tracks, depending on the location and population density.

One example of knotted train movements caused by a 100-year-old layout is Harold Interlocking, a 2-mile (3.2-km) complex in Queens, New York. Harold Interlocking is adjacent to Sunnyside Yard and just east of the East River tunnels. The interlocking is nearly identical to its 1908 layout (Figure 5.12).

On the west end of Harold Interlocking are four tracks from the four tunnels crossing the East River from Pennsylvania Station (marked with a "T" on Figure 5.12). During the morning rush hour, all eight tracks converge into four tunnels, and in the evening the process reverses. During rush hour, more than 40 trains per hour are sorted by Harold Interlocking.

A mechanical interlocking machine operated Harold Interlocking until 1990. Twenty-eight hundred relays were needed to duplicate and auto-

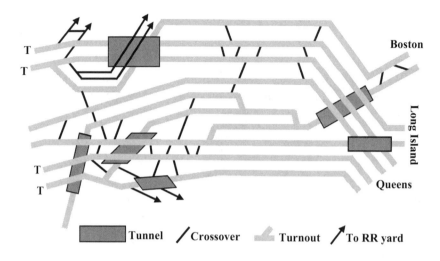

FIGURE 5.12. Harold Interlocking is located in Queens, New York, just across the East River. This interlocking sorts out commuter and Amtrak trains going to and from Pennsylvania Station. "T" indicates tracks coming from the East River tunnels.

mate the operators and mechanical locking bed controlling the switches, signals, remote signals, verification circuits, locking, and interlocking.

All interlockings on the Northwest Corridor were reconfigured to speed up passage of the high-speed passenger trains. Abandoned tracks and tracks in disrepair caused nonsensical switching patterns. For example, passenger trains had to slow and switch tracks just to slow and switch back all within the same interlock.

Reconfigured interlocks involved eliminating track crossovers for passenger trains or redesigning the interlocking so that crossovers could occur at higher speeds—up to 80 mph (129 km/h). A high-speed train forced to negotiate 124 automated interlockings is reason enough to explain the problems of high-speed rail on the Northeast Corridor.

Europe and Japan (and an increasing number of other countries) build new high-speed rail without interlockings. The new tracks are dedicated to the high-speed trains and are not shared with slower-speed commuter trains. They do not have to contend with 100-year-old tunnels and bridges and curves laid out for train speeds in 1850. The Northeast Corridor is a nineteenth-century system adopted for twenty-first-century use.

Another nineteenth-century problem involves road crossings. Unlike other high-speed trains around the world, the NEC is actually crossed by automotive traffic—a serious safety hazard. By 1986, all 34 road crossings between Washington, D.C., and New Haven had been removed. From New Haven to Boston, 11 road crossings remain, including 5 mandated by Congress.[11] Currently, the 150-mph (241-km/h) Acela slows down to 100 mph (161 km/h) at all road crossings.

CENTRALIZED TRAFFIC CONTROL

With electric relay systems the hand levers that operated the mechanical locking bed could finally be eliminated and the interlockings could be remotely controlled. Remote-controlled signals and interlocks led to centralized traffic control (CTC). The train crew drives the train following signals and/or instructions from the dispatcher. CTC reduces human error and the number of signalmen at interlocks.

The first "running meet" is believed to have occurred on a CTC track in 1927. In a running meet, one train is routed off the main track and onto a siding so that the two trains pass each other in opposite directions without stopping.

Today the busiest, most important main line tracks, including the tracks involved in the September 12, 2008, Los Angeles collision (see Chapter 2), are controlled by centralized traffic control.[12]

The railroad industry can increase throughput by adding more track or by using a better signal system. A common problem is operating 79-mph (127-km/h) passenger trains and 40-mph (64-km/h) freight trains on the same tracks. Coordinating meets and passes becomes problematic. With better traffic control the trains travel at higher speeds, are spaced closer, have running meets, and spend less time waiting to pass each other.

When a double-track system needed new ties and rail, it was more cost-effective to upgrade one of the tracks with CTC and use part of the second track for sidings. This reduced maintenance costs (and real estate taxes) while increasing throughput. In 1957, New York Central installed centralized traffic control on the 163 miles (262 km) of routes between Buffalo and Cleveland. Four tracks were replaced with 2 CTC tracks with crossovers every 7 miles (11 km). The entire stretch was controlled by just

2 dispatchers sitting in front of their control panels. Freight train speeds were increased from 30 to 60 mph (from 48 to 97 km/h). Automatic train stops (discussed later) were also added. At the time New York Central boldly stated this was the safest route in the world.

Often one track upgraded with CTC replaced two tracks without CTC. The standard centralized track control installation consists of a single main line track with sidings spaced perhaps every 25 miles (40 km) for lightly traveled routes and up to every 5 miles (8 km) for busier track. Interlockings are installed at the ends of sidings to control train meets. Typically, block signals are used to safely space trains in between sidings.

Computer analysis of actual train movements in the late 1990s concluded practical, safe limits for two-way traffic on a single track in dark territory, a single track with automatic block signals, and a single track with CTC as 15 MGT (million gross tons per year), 35 MGT, and 75 MGT respectively.[13] In all cases sidings were available for train meets.

A traditional CTC control board (Figure 5.13) consisted of a schematic of the track with lights showing the location of track switches, signals, and trains. Underneath were two rows of toggles and a row of buttons. The dispatcher set the correct route through the interlock by first setting the track switches and then the track signals. The system logic would not allow settings that collided trains. The buttons started a sequence of events corresponding to the levers' settings. Banks of relays in the control board verified the current position of signals and switches in the field, made the required changes, and verified that the changes had, in fact, taken place. A typical track switch motor required 7 seconds to unlock, move the switch, and relock.

In 1937, a technological advancement reduced the dispatcher's task to pushing only two buttons: the train's entrance and exit points through the track network. After the entrance point was selected, the control panel would light up all possible safe routes allowed by the relay circuit logic. The dispatcher selected the final route by pushing the button identifying the correct exit point. The relay circuits would automatically check for conflicting train routes and correctly set all the track switches and signals between the entrance and exit points. The push button entrance-

FIGURE 5.13. Active Union Switch and Signal CTC machine at Amtrak's THORN Tower. Although many of these systems still exist, a computer screen is more common today.

exit technology remains the standard today. More recently, the computer allows the dispatcher to program multiple trains in sequence.

Traditionally, dispatching was done from one or two desks with dispatchers communicating by telegraph with signalmen located near the tracks. As electronic controls and communications improved, the CTC centers slowly grew into the computerized centers they are today.

Mergers and computerization encouraged larger regional and eventually centralized dispatching for the entire train network. Today BNSF maintains the largest centralized dispatching facility in the United States, employing 533 dispatchers to cover 98 desks in its Network Operations Center located in Fort Worth, Texas. BSNF operates 6,400 locomotives on about 32,000 route miles (51,500 km) with 50,000 track miles (80,500 km). (Two miles of triple track is 2 miles of route and 6 miles of track. Most routes are single tracked.)

THE DISPATCHER

Whether in signaled or dark territory, all train traffic is controlled by a dispatcher. Nothing moves without the dispatcher's permission. The dispatcher grants authority to enter the main track, plans train meets, allows higher-priority trains to pass slower trains, and sorts out trains crossing at junctions.

The level of responsibility (and associated job stress) is very obvious— trains are commonly routed for head-on, rear-end, or crossing collisions. The dispatcher's job is to keep the trains moving at a safe distance from each other. A routine slowly develops as new trains enter the dispatcher's territory at about the same rate as others leave. Everything must be adjusted, however, when unexpected delays occur.

Equipment, track, signal, and communication failures can all cause delays, as can weather, vandalism, crew changes, switching in the yard, track obstructions, and, of course, accidents. Trains must be routed around maintenance crews required to inspect and repair track—both activities with their own unexpected delays.

The dispatcher must monitor the location of all trains, track inspectors, maintenance crews, and hazardous material. The dispatcher must adjust for different train speeds, the train crew's hours of work (12 hours maximum), location of relief crews, and the weather; even the number of locomotives and number of cars will alter train speeds up and down steep grades.

Even ordinary train movements can require extensive activity for the dispatcher. In Chapter 9 four locomotives repeatedly go back and forth and on and off the main track to pick up their train, reposition the locomotives at the head of the train, move off the main track for oncoming traffic, and couple to their helper locomotives. After 5 hours, the train has advanced only 19 miles (30 km).

Computer software helps the dispatcher keep track of everything, even in dark territory—but this help is somewhat muted because he or she is then expected to supervise more trains.

Despite the responsibilities of the job, dispatcher-caused accidents are rare. A 1995 FRA study found only five reportable accidents caused by dispatchers during a three-year period. Even then, in four of the five accidents the train crew incorrectly repeated their instructions, and the dispatcher failed to notice.

Any potential dispatcher error should be flagged by a computer. Information is continuously presented to the dispatcher, who should be able to correct any mistake at a later time. For the train crew, missing a signal during minutes (or even seconds) of distraction is not as easily corrected.

CENTRALIZED TRAFFIC CONTROL ON THE NORTHEAST CORRIDOR

On the Northeast Corridor the trains are powered by electricity instead of diesel engines. Electrification played out differently in America than in the rest of the world. Up until World War II, the electric NEC was the most advanced railroad system in the world. After World War II, dieselization took over in the United States. Prosperous Americans loved their cars and spread out into suburbia with their new federally built highway system. The bombed-out economies of Europe and Japan could not afford millions of cars—their people were going to move by train or not at all. American electric train technology fell by the wayside and the Northeast Corridor fell into disrepair until modernized with today's dominant foreign technology.

Electricity is the most efficient way to operate trains. Even in 1914, when the first 70-mile (113-km) stretch was electrified, burning coal in a power plant created twice the horsepower generated by burning coal in a steam locomotive firebox. A large pile of burning coal with a huge chimney at a power plant can burn hotter and more efficiently than a small pile inside a locomotive. A steam locomotive consumes coal even when idling or coasting. An electric motor has less friction than all the moving parts in a steam or diesel engine. Also, an electric train does not have to carry around fuel, a fuel tank, an engine, and a bigger frame for all the added weight. Today all high-speed trains in the world operating above 125 mph (201 km/h) are electric for these reasons and others. The initial costs, however, are roughly double for the added electric infrastructure.

Once the travel time between two destinations comes down to three hours or so, all over the world trains compete successfully with planes and dominate the market. Profitability, especially in the case of high-speed electric trains, requires a minimum population density to work. Such population concentrations exist in Japan, Europe, and the NEC in America.[14]

IN ADDITION TO MOVING TRAINS AROUND, dispatchers on the Northeast Corridor must also move electricity around. On the NEC traffic control is called Centralized Electrification Traffic Control, or CETC. CETC controls the trains and regulates the flow of electric power, balances the loads, shuts down sections for maintenance, and monitors electric switches, circuit breakers, transformers, and other power controls.

This is no trivial matter. With more than 1,800 trains operating on the NEC per day, mostly concentrated at rush hour, power demands are constantly changing. Depending on the time of day and the number of trains on the tracks, electric power is constantly being turned on and off, adjusted, and balanced in the network with electric switches in the interlocking towers.

Most of the Northeast Corridor has the same electrical layout as it did when first electrified in the 1930s. In fact, quite a bit of the original electrical equipment has survived into the twenty-first century.

Alternating current (AC) is more efficiently transmitted over long distances at high voltages. Since electrical resistance is fixed, increasing voltage results in lower current (per Ohm's Law, voltage = current × resistance). Reduced current decreases the power losses caused by the flowing electrons rubbing on each other.

On the Northeast Corridor power plants supply 13,200 volts increased to 132,000 volts with transformers for more efficient transmission. The voltage is then decreased with more transformers to 11,000 volts in 67 substations every 7–10 miles (11–16 km) for the conductor wire. The pantograph rubs on the conductor wire and supplies power to the train. Additional transformers on the trains further step down the voltage from 11,000 to the 625 volts used by the motors.[15] For safety, only 36 volts, supplied from batteries, is used inside the crew compartment for signal equipment.

The 132,000 volts is not supplied directly to the conductor wire because arcing will damage the pantograph and/or electric switches. Arcing is a high-temperature electron discharge across a gap. If the contacts of a switch (or between the pantograph and conductor wire) are separated by a small gap, the electricity will jump the gap like a tiny bolt of lightning, with temperatures as high as 9,000°F (4,982°C). Electric switches are required to isolate sections for maintenance. Each track has four redundant conductor wires switched on and off as needed; an entire substation can be shut off and bypassed.

Today the electric power loads on the Northeast Corridor are balanced with computers. In the 1930s and for many years afterward, current, voltage, and power calculations were solved with a mini electric panel that mimicked the real thing. Any adjustment to the big grid would first be tested on a mini panel located in the Central Load Dispatching Office in Philadelphia. After any load change, acceptable current and voltage levels were checked on the mini panel with portable meters.

THROUGHOUT THE 1980S, the Northeast Corridor was being modernized in preparation for high-speed trains. This was and remains a massive project. The rails had to be upgraded and the roadbed stabilized. Hundreds of 100-year-old bridges and a few tunnels had to be brought up to modern safety standards or replaced, and the signals had to be updated.

In 1975, just 25% of the track was signaled for traffic in both directions. Eventually, the Northeast Corridor evolved into mostly three tracks signaled for bidirectional movement. All the interlocking logic had to be redesigned. In 1987, the interlocking control towers were slowly being automated and connected to a central dispatcher. Beginning with a 110-mile (177.02-km) stretch from Philadelphia to Wilmington, Delaware, in the spring of 1987, Centralized Electrification Traffic Control was gradually introduced on the NEC. The 124 manually operating interlockings slowly became modernized, automated, and connected to 1 of 3 control centers (Philadelphia, New York City, or Boston). Today the NEC is controlled by 19 dispatchers in 3 control centers aided by an occasional operator in a tower.

PEAK COLLISION PERIOD

In the early part of the twentieth century, the collision rate reached its zenith. This period also coincides with record railroad passenger traffic. Between 1906 and 1921, there were 26,297 head-on or rear-end collisions, resulting in 4,326 fatalities—almost 1,400 collisions and 270 fatalities per year. The average passenger fatality rate for this period was about 2.2 fatalities per 100 million passenger miles traveled. Contemporary rates today are 0.034 per 100 million for intercity travel and 0.02 per 100 million for commuter trains.

EARLY TRAIN CONTROL

Automatic controls to prevent collisions began more than 100 years ago. Mechanical train stops were first used on Boston transit trains in 1901. A mechanical trip would trigger a valve on the train that activated the emergency airbrake system.

Somewhat surprisingly, mechanical stops remain in use on the New York, Chicago, Boston, Philadelphia, and Toronto transit systems. (As of 2004, the New York Transit subway system's 856 miles [1,378 km] had 11,646 blocks protected by track circuits and automatic train stops.) A short mechanical arm, called a trip stop, is pneumatically or electrically raised beside the rail when a signal shows the next block is occupied. The trip stop strikes a trip cock on the train that evacuates the air in the brake line and triggers emergency braking.

Passenger trains began experimenting with mechanical train stops and installed more than 100 miles (161 km) on 3 stretches between 1913 and 1919. Because of wear and track-shifting under heavy freight trains, mechanical stops for freight trains were less effective.

Electric automatic train stops came next. An alternate current in a coil creates an alternating magnetic field. The changing magnetic field will "induce" (without contact) an AC current in a second coil attached to the locomotive, the basic principle of a transformer. This AC current, now inside the locomotive, can be used to activate the brakes.

In 1922, after extensive studies and congressional hearings, the ICC ordered 49 railroad companies (expanded to 93 railroads in 1925) to install automatic train stops (ATSs) or other safety devices on at least one major line by the end of 1925. The railroad companies resisted this order furiously, with much justification. It was impossible back then to design a device that safely stopped long and heavy freight trains and light passenger trains. Braking is not a precise process and varies greatly with the condition of the rails and the length and weight of the train. Chapter 6 explores the difficulties of precise stopping even with modern technology. Also, a sudden emergency stop can derail a heavy freight train.

Because the ICC regulated prices and would usually not allow increases, the railroads had trouble justifying any extra costs unless it improved operations. Forced stops, when merely slowing would suffice, played havoc with schedules. After emergency braking depleted the air, it could take up to 45 minutes to re-pressurize the airbrakes and many more minutes regaining lost speed. At that time even the Brotherhood of Locomotive Engineers (the engineers' union) argued that automatic devices would decrease safety as the engineers became over-reliant on them and paid less attention.

These arguments went on for years. Meanwhile, the number of passenger deaths dropped significantly (mostly from reduced ridership) while fatalities at road crossings tripled between 1923 and 1927. In 1928, the ICC stopped demanding train controls on additional routes, concluding that more lives could be saved by spending more money on signals at road crossings.

The Pennsylvania Railroad (PRR), always a technological leader, began widespread use of automatic cab signals. Automatic cab signals moved the trackside signals (called wayside signals) inside the locomotive cab. This was done by sending pulses of current into the rails. The pulses could be displayed as signals inside the cab. With cab signals now located on the engineer's control panel (with a whistle announcing changes) it was extremely difficult for a conscious engineer to miss a signal even during inclement weather. The PRR demonstrated that cab signals were just as safe as train controls, and the ICC allowed Union Pacific, PRR, and others to substitute cab signals for automatic train stops in the late 1920s.

Actually, it is a far more confusing story. After ATS or ACS was added, it was not necessarily easy to get permission for removal, and remnants of these safety features still exist today. In general, the ICC allowed removal of ATS and ACS when passenger service ended on that route.

SPEEDY PASSENGER TRAINS IN THE 1930S

Speed limits eventually placed on the streamliners introduced in the 1930s are an important part of the train control and signal story. The same rules limit passenger train speed today.

The first official record of more than 100 mph (161 km/h) occurred in 1892, when a U.S. steam-powered train averaged 112.5 mph (181 km/h) for 1 mile. The record was broken in 1903 with a brief burst of 124 mph (199 km/h). Short runs are less meaningful than sustained longer runs. For this reason an international railway study identified speed records for 15 distances ranging from less than 25 miles (40 km) to greater than 3,000 miles (4,828 km).

Speed records began falling in the mid-1930s with the introduction of diesel-powered high-speed trains, the Burlington's Zephyr being the first of many. In 1936, the United States held the record in 12 of the 15 categories and dominated until Japan developed Bullet trains in the 1960s. (The current record for a conventional passenger train is 357.2 mph [575 km/h], made during an experimental test run on a modified TGV French passenger train in 2007.)

The Burlington lost 60% of its passengers between 1926 and 1931. Besides the Great Depression, one of the reasons was, of course, the automobile. Starting in 1901 with 15,000 cars, the number increased 10 times every 7 years. By 1933, there were 55,000 locomotives pulling nearly 52,000 passenger cars trying to compete with 25 million cars.

Burlington's innovative (and risky) plan was to build a lighter, faster, stylish passenger train. The Zephyr, with its art deco design, had many innovative "firsts": the first aerodynamic streamlined design, the first high-speed train powered by a diesel engine, and the first lightweight train—about half the normal weight. Wind tunnel testing showed a streamliner could reach 100 mph (161 km/h), whereas a conventional train with the same horsepower could only obtain 75 mph (121 km/h).[16]

On May 26, 1934, the Zephyr was introduced to the public on its publicity-filled, record-breaking run. Although only briefly hitting 115.2 mph (185 km/h), the Zephyr reduced the travel time between Denver and Chicago from 25 to 13 hours and averaged 77.75 mph (125 km/h) for 1,017 miles (1,636 km)—the longest nonstop train ride ever. Just 2 years later, the same run was completed at 83.3 mph (134 km/h).

Because of the intense public attention on the Zephyr's maiden trip, Burlington went to extraordinary means to avoid any mishaps. All other trains were parked along the entire route, and flagmen protected every road crossing. To prevent tampering, track switches were spiked shut.

Speed limits were posted on difficult curves. Today modern technology creates similar protection for Amtrak's high-speed Acela.

The Zephyr was a sensation wherever it appeared. In 1934, the Zephyr starred in its own Hollywood movie, *The Silver Streak*. The Zephyr, the "hero" of the movie, raced across the continent transporting an iron lung for a medical emergency.

The new schedule between Minneapolis and Chicago had diesels making the trip 3.5 hours faster than steam engines and 115 minutes faster than Amtrak trains today. In 1935, revenue doubled for the Zephyr and operating costs were about half. By 1938, 90 streamliners were in service across the nation.

THE RULES CHANGE AGAIN

On April 25, 1946, two passenger trains collided, killing 45. A trainmaster on the Advance Flyer thought he saw something dragging under the train. Since dragging equipment can derail a train, the trainmaster promptly ordered the Advance Flyer stopped 28 miles (45 km) southwest of Chicago. Unfortunately, 90 seconds later the Exposition Flyer, traveling at about 75 mph (121 km/h), collided into the stopped Advanced Flyer (Figure 5.14).

A signal light 1,100 feet (335 m) east of the stopped train indicated red for "stop." A second signal 5,150 feet (1,570 km) farther indicated yellow for "slow down." The investigators determined that the yellow light was visible nearly 2 miles (3.21 km) from the stopped train. The engineer of the colliding train stated that he saw both lights, but at 85 mph (137 km/h) the train was traveling too fast to stop before hitting the Advance Flyer. The investigation did not favor the colliding engineer's story.

Several test runs were conducted to determine the train's stopping

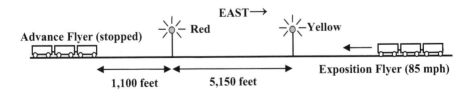

FIGURE 5.14. Collision on April 25, 1947.

distance. Representing the shortest possible warning consistent with the engineer's story, it was possible the yellow warning light came on just as the Exposition Flyer passed the signal. Braking for the third test began at 80 mph (129 km/h) at the yellow signal light. The test train stopped in 5,191 feet (1,582 m), still 1,329 feet (405 m) short of the collision site. Repeating this test at 85 mph (137 km/h) the test train took an additional 360 feet (110 m) to stop. Emergency brakes were used during the final test. The test train stopped in a smoky, sparking 3,300 feet (1,005 m). As a result of this particularly bad collision, the ICC altered the haphazard rules first made in 1922 (and frequently waived) concerning automatic train controls.

IN 1947, the ICC passed new rules requiring automatic block signals for all freight trains operating above 49 mph (79 km/h) and passenger trains operating above 59 mph (95 km/h). This rule affected about 18,000 miles (28,968 km) of track. Additionally, all passenger trains operating above 79 mph (127 km/h) (about 27,000 miles [43,452 km] of track) now required train controls (i.e., automatic train stops or automatic cab signals). These rules are still in place today.

The ICC expected expanded use of train controls. Although train controls were added to about 5,000 miles (8,046 km) of track, most railroads, in the face of declining passengers, simply lowered operating speeds to comply. The decline of passenger trains in the United States accelerated. The number of intercity trains plunged from 20,000 in 1929, to 2,500 in 1954, and to about 450 in 1970, when Amtrak took over. Today in the United States high-speed trains are being studied for many locations, but only a few passenger trains currently operate above 79 mph (127 km/h). The Northeast Corridor is the most important example.

In 1951, the Pennsylvania Railroad continued innovating and developed automatic train control (ATC) to regulate train speeds. It added ATC to passenger trains already equipped with automatic cab signals. ATC took the electric pulses sent in the rail for automatic cab signals one step further. The coded track pulses creating signals inside the cab were developed to apply the brakes if the engineer did not respond to a warning light and whistle within 6 to 8 seconds. Speeds of 20, 30, and 45 mph (32, 48, and 72 km/h) were enforced. Today ATC on the Northeast Cor-

TABLE 5.2
Federal Railroad Administration Surveys of Train Control Use,
1993 and 2007

	Total Track Miles (1993)	Total Track Miles (2007)
ACS	3,964	5,303
ATS	1,780	2,095
ACS/ATS	90	310
ACS/ATC	1,554	2,106
ACS/ATC/ATS	2,455	419
Total	9,843	10,233

ridor has evolved into a more advanced system known as Positive Train Control (PTC; see Chapter 6).

Automatic train control differs from automatic train stops. Automatic train control enforces the speed limit but will not precisely stop the train because the signal is received randomly in a long section of track. Also, it takes 3 to 5 seconds for the changed signal to appear in the cab, and the engineer is allowed 6 to 8 seconds to respond. ATC can enforce any speed except zero. The slowest speed enforced is 20 mph (32 km/h). Trains with ATC can still travel past a signal at speeds below 20 mph.

With ATS the engineer has 6 seconds to acknowledge the alarm before the train's brakes are applied. Because braking is not a precise and re-peatable process (especially on heavy freight trains), the designers allow overriding of the automatic stop by merely pressing a button to acknowl-edge the alarm. This gives final judgment to the engineer (to account for all the variability associated with a long, heavy freight train) and reintro-duces the possibility of human error.

Despite the near disappearance of fast trains, remnants of train con-trols still remain (Table 5.2). ATS, ATC, and ACS (automatic cab signals) are often used in combination.

TRAINS PROTECTED BY CTC, ABS, ACS, AND ATS COLLIDE

On July 2, 1997, two Union Pacific freight trains collided in spite of both being protected by a centralized traffic control system, au-tomatic cab signals, and automatic train stops.

This busy section of Kansas main line track averaged 60 to 75 trains per day. Train 1 passed a signal indicating "do not exceed 40 mph (64 km/h)." Just 2.4 miles (3.8 km) later, a second signal warned Train 1 to slow to 30 mph (48 km/h). The dispatcher remotely threw a switch directing Train 1 onto a siding to allow a second oncoming train to pass. After the 6,400-foot (1,951-m)-long train entered the 13,500-foot (4,115-m)-long siding, a third signal warned Train 1 to stop at the next signal. Two-and-a-half (4 km) miles later, at the end of the siding, a fourth signal ordered Train 1 to stop and wait for oncoming Train 2 to pass on the main track.

The collision occurred at 2:23 a.m. and killed the engineer of Train 1. The investigators concluded that the engineer fell asleep somewhere in the siding. The engineer was sufficiently awake to acknowledge the alarm from the automatic cab signals after each signal changed. He did not slow the train, however, when entering the siding.

Even though the automatic train stop did activate the train's brakes, the train was moving too fast to stop. The sleeping engineer woke up when the brakes were automatically applied. Exact speeds were not recorded because the event recorder was destroyed in the collision and fire.

Operating rules for these two freight trains (and all trains) require the engineer to call out all signals to the conductor. The conductor, sick in the lavatory for 20 minutes, missed all callouts. The colliding train struck the sixth car of the second train, derailing 15 cars. The subsequent fire engulfed the derailed cars and locomotives of both trains.

The focus of the 1947 Chicago collision investigation was completely different from the one in this 1997 accident. In 1947, the investigation centered on proving that the train could have easily stopped and concluded that the engineer missed the yellow and red signals. Clearly, the engineer missed the signals in the 1997 accident. In a relatively modern development, the investigators sympathetically view the human potential for error and do not even assess safe stopping distances.[17] In the case of this accident, the investigators focused on the engineer's sleep patterns, changing schedules, "the rhythmic sound and motion produced by the locomotives,"[18] the lack of sensory stimulation inside the cab at night, and Union Pacific's existing operator fatigue education program.

The National Transportation Safety Board also recommended developing Positive Train Control, designed to prevent such accidents. PTC is the subject of the following chapter.

6

Positive Train Control

Perhaps the most violent collision in U.S. history occurred in Chase, Maryland, on January 4, 1987. At 1:16 p.m. an Amtrak train with 2 locomotives, 12 cars, and 660 passengers left Baltimore's Penn Station on track 2 on the Northeast Corridor. At about the same time 4 miles (6.4 km) north, 3 Conrail locomotives departed on track 1.

Just 16 miles (26 km) north of Penn Station is Gunpow Interlocking, a controlled interchange for merging train tracks. At Gunpow the four main tracks converge to two before crossing the bridge over the Gunpowder River.

The dispatcher's plan was for the Conrail locomotives to stop at Gunpow and wait for the passenger train to pass. Instead, the Conrail locomotives drove past their stop signal and onto track 2, just 3,000 feet (914 m) in front of the 120- to 125-mph (193- to 201-km/h) Amtrak train. The collision occurred at 1:30 p.m., just seconds after Conrail locomotives entered track 2.

The rear-end Conrail locomotive exploded into a fireball and disintegrated. Downed power lines ignited spilled diesel fuel. The fire department arrived 7 minutes after the collision from a station just 1.4 miles (2.25 km) away. Fire was not a survival factor.

Both Amtrak locomotives and the first three passenger cars were destroyed. Except for the first Conrail locomotive (shoved 900 feet [274 m]), all other cars and locomotives derailed. Fortunately, the first car was an unoccupied food service car. Thirteen of the 25 passengers in the second car were killed, as well as the Amtrak engineer and two passengers in the third car.

The Conrail locomotives had multiple warnings to slow and stop. The first signal light, visible 2 miles (3.2 km) from the crossover, warned the locomotives to slow to 30 mph (48 km/h) and prepare to stop at the next signal. Passing the first signal activated Automatic Cab Signals (ACS) on their control panel. The automatic cab signal repeats the track signal inside the cab, making it far more difficult to ignore. If the crew does not acknowledge the changing cab signal by pressing a button, a warning whistle sounds. The shrill 95- to 105-decibel whistle, however, had been silenced with duct tape.

A second light, visible more than a mile (1.6 km) from the crossover, signaled the crew to stop. A second cab signal, almost a mile from the track switch, required slowing to 20 mph (32 km/h). The silenced whistle again tried to sound an alarm.

The 3 locomotives were moving at 64 mph (103 km/h) when the train crew realized their error and applied their emergency brakes 2,678 feet (816 m) from the point of collision. The 3 locomotives finally stopped on the wrong track just past the switch. Investigators concluded that the locomotives could have stopped short of the crossover if traveling 55 mph (89 km/h) or less. The normal speed limit was 50 mph (80 km/h).

Testing after the accident demonstrated that an alert crew traveling at 65 mph (105 km/h) could have stopped in about 2,500 feet (762 m) (2,200 feet [670 m] before the track switch) by responding when the stop signal first became visible.

The automatic signal lights on track 2 detected the errant freight train and signaled the Amtrak train to stop. The earliest moment the Amtrak engineer could have noticed this signal was about 3,000 feet (914 m) from the collision point. After emergency braking, the Amtrak train slowed to 107 mph (172 km/h) before impact.

Assuming the engineer took 3 seconds to notice and respond to the signal,[1] there was about 2,500 feet (762 m) to stop the train. Not counting operator reaction time, computer simulation predicted stopping in

6,470 feet (1.972 m) from 122 mph (196 km/h) at full service braking and 4,382 feet (1,336 m) during emergency braking.[2]

Each Amtrak locomotive weighed 202,000 lbs (91,625 kg) and produced 7,000 horsepower (5,220 kW). The Amtrak train weighed 836 tons (758 MT). The three Conrail locomotives weighed 405 total tons (367 MT).

The Conrail conductor testified he had been busy making lunch. Evidence later showed the Conrail crew were heavy and frequent users of marijuana. The engineer was convicted of manslaughter by locomotive and received a prison sentence of five years.

In 1987, automatic train control (ATC) enforced speed limits on all passenger trains on the Northeast Corridor but not on freight trains. ATC did not exist on freight trains because of the problems associated with stopping long freight trains (described in Chapter 5). Stopping trains accurately, even with modern computer controls, remains a problem. Recall that the ATC signal is received randomly within a track block and cannot be used to enforce a stop at a specific location. For that reason ATC does not enforce zero mph.

A new and improved ATC was added to freight trains on the Northeast Corridor after the 1987 collision. Because ATC does not enforce stops, the freight train still could have driven in front of the passenger train at speeds below 20 mph (32 km/h). If a 20 mph speed limit had been enforced, presumably the Conrail locomotives would have been able to stop in time before accidentally changing tracks.

To operate the new Acela at 150 mph (241 km/h) on the Northeast Corridor the Federal Railroad Administration (FRA) required an additional safety system known as the Advanced Civil Speed Enforcement System (ACSES). If an engineer tries to drive past a stop signal at any speed, ACSES takes control of the train and forces a stop. ACSES was installed in 2000. Also added were additional speed controls. The speeds now enforced are 0, 20, 30, 45, 60, 80, 100, 125, and 150 mph (0, 32, 48, 97, 129, 161, 201, and 214 km/h).

To stop trains transponders were added to the existing automatic train control signals. The transponders stop the train at specific locations and enforce speed limits on curves. The transponders work as follows.

Antennas under the locomotive continuously transmit a signal into the ground. This signal powers the transponders mounted between the rails. When a transponder is energized with radiated power, it transmits an encoded message. The message includes the distance to a stop (or the beginning and end of speed reduction for a curve). Each transponder also contains information about any upcoming hills that will affect braking.

The locomotive's receiver captures the transponder's message and turns the information over to the locomotive's computer. The onboard computer calculates required braking. The calculation considers the type of train (i.e., freight train versus different types of passenger trains), the train's speed, distance to the speed change, and changes in elevation. Each locomotive is also equipped with an axle rotation counter or tachometer to measure speed and distance traveled—information needed for the calculations.

The old automatic train stop systems (ATS), described in Chapter 5, also used transponders, albeit "dumb" ones only capable of activating the brakes. The ACSES transponder is "smart" by comparison, containing unique information at each location. The ACSES system guides the driver on how to drive the train, warns the engineer about speed violations, and applies the brakes automatically if needed. The dispatcher can also program into the system temporary "slow orders" to protect track workers.

Amtrak's ACSES became the first system to meet all the requirements of a fully functional Positive Train Control (PTC) in the United States. This new PTC system would have prevented the 1987 collision by stopping the Conrail locomotives after they ignored their stop signals.

On the Northeast Corridor the existing ACS and ACT in use since the 1950s was considered to meet 90% of the requirements of PTC. The gap was closed by adding the transponders previously described. The transponders told the train where and how to stop. Elsewhere, without an extensive signal system in place, wireless PTC is being developed.

POSITIVE TRAIN CONTROL

Positive Train Control (PTC) is the biggest safety improvement in modern railroading. PTC is new technology that promises to prevent train accidents caused by human error. If the human misses a signal to slow or stop the train, the PTC system takes over the train and responds as needed. Specifically, PTC is designed to prevent collisions, enforce speed limits on curves, check switch status, and protect work crews from errant trains.

The Positive Train Control system consists of advanced communication systems that continuously monitor the train's position and authority,[3] digitally display the track on the engineer's monitor (including hills, road crossings, track signals, location of work crews, speed limits, trackside defect detectors, etc.), and monitor the throttle-computer and brake-computer interfaces. The PTC system will automatically intervene if the train crew exceeds the speed limit, misses a signal to stop, or the humans become incapacitated.

Positive Train Technology is actually a generic term referring to a host of specific technologies being developed.[4] The dominant technology emerging uses a global positioning system (GPS) to locate trains and wireless digital communications to control and stop trains as needed.

The first serious attempt at modern Positive Train Control began in 1983. In 1987, Union Pacific announced aggressive plans to expand its experimental PTC system to its entire network within five years. The system worked, but failsafe design methods required the system to return to a safe state (i.e., stop the train) if anything failed. The 1980s radio data link was similar to the internet dial-up modem of the same period and could not keep up with the required information flow. The trains simply stopped too often for efficient operations.

Meanwhile, the FRA formed a committee of experts in 1997 to recommend new rules to address the latest technological developments. The committee consisted of railroads, suppliers, labor unions, and government agencies. Rules for voluntary adoption of PTC were first written in 2005.

Based on established safety analysis, the FRA reported to Congress in 2004 and 2009 that PTC was not cost-effective. Companies of course have to watch their bottom line and are never anxious to spend an extra few billion dollars on unproven new technology. Commuter trains, owned by regional government agencies, reached the same conclusion. In fact, in October 2009, Amtrak told the FRA it feared that installing PTC would put the company out of business.[5]

How did the FRA conclude PTC was not cost-effective? The FRA and many other federal agencies, including the Federal Aviation Administration (FAA), evaluate safety improvements by weighing the cost versus any perceived benefit. If the FAA or FRA determines that new safety features are cost-effective, the federal agency mandates the safety improvements.

The projected costs of new safety equipment are usually straightforward. The benefits are a bit more nebulous. The FRA (and FAA) estimates the reduction of accidents over a 20-year period and associated savings. Potential savings is projected value of damaged equipment and people. Using $6 million for the price of a human, the benefits were projected to be $1 for every $20 spent.[6] The FRA did not mandate PTC technology.

Congress intervened and took the decision-making process out of the hands of the FRA by passing the U.S. Rail Safety Improvement Act of 2008 (RSIA). RSIA requires PTC systems installed by 2015 on all passenger trains, freight train routes with more than 5 million gross tons (MGT) of freight per year, and routes that transport hazardous materials. The rule is expected to affect about 73,000 miles (117,500 km) of the national rail network, about 45% of the total. Final technical rules were issued by the FRA in early 2010. (Negotiations in 2011 reduced the miles of affected track by about 10,000 miles.)

Positive Train Control will not prevent all collisions. If a train derails onto a nearby parallel track, nothing currently proposed prevents a second train from colliding with the derailed train. Also, existing track signals often permit a following train to enter occupied track at reduced speeds. PTC will enforce the speed reduction, but will not actually prevent slow-speed collisions.[7] Also, a train on tracks without PTC can still collide with a protected train where the tracks cross. Derailments caused by equipment failures will also not be prevented. Main track derailments are typically 15 times more likely than main track collisions.

IT IS IMPOSSIBLE to keep up with the rapid developments in PTC technology. Instead, the rest of this chapter will address the simple question of why modern technology can't prevent train accidents. It can and will—it just hasn't been as easy as one might think. Regardless of the 2015 deadlines, PTC will remain a work in progress for many years because of the problems described here.

WHY CAN'T COMPUTERS CONTROL TRAINS?

For decades metropolitan rapid transit systems have been computerized. Why hasn't the national rail system followed suit? Besides the obvious answer of cost, there have been many technical difficulties.

Before we consider the Washington Metropolitan Area Transit Authority and some of its problems with computer controls, a few definitions are in order.

Railway versus Transit Operations

Railways operate on the national rail system, made up of about 162,000 miles (260,700 km) of connected track.[8] Fundamentally, any train can reach any location on the network. Railway operations include intercity trains, commuter trains, and freight trains—all sharing the same track. Unfortunately, trains operating at different speeds on the same track are more likely to collide.

Railways are regulated by the Federal Rail Administration. Railway equipment shares many similar design features. Transit systems, going by a variety of names (subway, metro, rapid transit), are regulated by the Federal Transit Administration and usually have custom-designed equipment.

Transit systems are fundamentally safer, as they should be. The number of people moved and the frequency of trains leaving the station create higher risk that demands more attention to safety. The Washington Metro, second to only New York's transit system, has more than 700,000 departures from 86 stations each weekday. Transit systems are safer because they operate on dedicated track. This means that they cannot collide with freight trains or road vehicles. Traveling at the same speed and in the same direction, transit cars are also less likely to collide with each other. The lighter-weight transit cars are easier to stop, have virtually no slack and handling problems, and are less likely to wear and damage the track. Transit systems usually have additional safety features similar to Positive Train Control being described here. Because transit systems are safer and because there are only 2,375 urban transit miles (3,822 km) of track, there are fewer accidents.

The Washington Metro

Washington's Metro has operated electrical trains controlled by computer since 1976. The computer controls the train's acceleration and braking, keeps the trains safely spaced, and positions the trains at the station platform. The entire network is overseen by the Operating Control Center. The Control Center coordinates train movements, electric power, communications, and station activity. The Control Center also maintains radio contact with the train operators and coordinates any anomalies (i.e., emergencies, track maintenance, etc.).

The trains can be operated by computer or by an operator. In manual mode the operator accelerates and brakes the train. There are two types of manual mode: completely manual (used when moving trains around for maintenance) and manual mode with automatic spacing protection. Metro's automatic spacing protection is similar to operating a freight train with Positive Train Control.

Metro operations have changed over the years. The trains operated in automatic mode for the first 20 years except when conditions were wet or icy. It was thought the computer would have trouble adjusting to slippery tracks. Also, each operator used manual mode at least once a week to stay in practice.

In 1995, Metro decided to change to full-time automatic operations. At the time it was believed that manual braking was creating a new problem—wheels with flat spots. If the brakes grip the wheels with too much force, the wheels can slide on the rails and wear flat spots on the wheels. The flat spot pounds the rail with every rotation. A noisy ride, with accelerated wheel and rail wear, results.

On January 6, 1996, a Metro train operating in automatic mode failed to stop at a station, continued for another 470 feet (143 m), and struck a stopped, empty train at 22–29 mph (35–47 km/h). The train operator died in the collision. The crash investigators concluded that operating 24 hours a day in automatic mode had been a "hasty decision based on insufficient information."[9] Besides, automatic operations did not eliminate flat spots on the wheels.

At the time of the accident the normal Washington Metro operating speed was 75 mph (121 km/h) except during slippery conditions. Because of snow, the accident train was operating at a reduced speed of 59 mph (95 km/h).

The first car of the four-car accident train overshot the station by one car. Instead of backing the train up, the train operator and Control Center decided to leave the doors closed on the first car during the station stop. Because the first car was outside the station limits, the train's computer controls did not receive the correct signal to reestablish the train's speed at 59 mph (95 km/h). Instead, the train defaulted to its normal speed of 75 mph (121 km/h). The train's operator realized the train was going too fast and called the Control Center for advice. Both had the authority and ability to intervene and lower the speed. Neither of them had the training to understand that the train could not safely stop on frosty tracks at 75 mph (121 km/h). The human-computer interface failed.

Beginning at 2,700 feet (823 m) from the station, the approach to each station has 4 electronic sensors that send a signal to the train's brakes. The sensors are designed to gradually stop the train at a precise location along the platform. The Metro's brakes also have anti-slip sensors to prevent wheel slip, similar to ABS (antilock braking system) in cars. When the sensors sense slip (a speed difference between the wheels), the brake clamping force is reduced. Typically, the brakes are rapidly applied and released (i.e., the brakes pulse), a process familiar to many car drivers.

A stopping distance less than 3,422 feet (1,043 m) was required to prevent the collision. Three out of four test stops after the accident failed to stop the train in time. The test stopping distances varied from 3,353 (1,022 m) to 4,588 feet (1,398 m). After the accident, the anti-skid braking software was improved for wet weather and the speed of all Metro trains was reduced to 59 mph (95 km/h).

The errors in computer logic that increased the train's speed and the failure of the humans to intervene, even though they knew the train was traveling too fast, are considered classic examples of failures of automatic train systems. The current rules for PTC systems require an extensive hazard analysis designed to identify such problems.

The problems of stopping a 4-car train weighing 148 tons (134 MT) with a sophisticated braking system during snowy conditions is most definitely related to the problems of stopping a 150-car, 15,000-ton (13,600-MT) freight train with Positive Train Control. Also, trying to maintain uniform braking performance for Metro's 1,100 cars spread over 106 miles (170 km) of track is far easier than managing 1.3 million freight cars (built over a period of decades) on 160,000 miles (257,500 km) of track.

There are no plans to convert the nation's 1.3 million freight cars (with 10.4 million wheels and brakes) to computer-controlled anti-skid brakes, nor are there any plans to assist the stopping process with four rail-mounted sensors at every location on the national rail network where freight trains might stop. Freight trains frequently stop on sidings to allow faster trains to pass or to get out of the way of oncoming trains. Most towns that existed in 1920 have a railroad turnoff that is still usable if, for example, a bearing overheats.

The worst accident ever on the Washington Metro occurred on June 22, 2009, when a train collided into the rear of a stopped train, killing nine. An automatic track circuit failed to detect the stopped train. The train was traveling at 59 mph (95 km/h); the operator hit the emergency brakes just 425 feet (129 m) before the collision. The accident occurred during rush hour with the train operating in automatic mode. After the accident, the entire system was switched to manual mode for 12 days. Similar to a nineteenth-century manual block system, each station had personnel posted, granting permission for trains to enter.

Recall the automatic block system described in Chapter 5. Electrically isolated sections of track, called blocks, are electrified with a current that detects the presence of a train's steel wheels. For the Washington Metro, a block can vary from 150 feet (46 m) to 0.5 mile (0.8 km). The Metro system is designed to keep the trains separated by at least two blocks. When the track circuit detects a train, signals are transmitted to automatically slow any following trains. If everything is working correctly, a following train closer than two blocks will be automatically stopped.

Under normal conditions, if a track circuit stops working, the Control Center operators will immediately see it on their screen. Additional circuits on adjacent track will automatically force approaching trains to stop before they reach the "dark" section of track. In this case, however, the broken circuit was fluttering on and off so fast that none of the safety systems were activated. Careful review of computer data after the accident determined that the troublesome circuit had been failing intermittently for more than a year.

There was a similar incident in 2005. A train entering a station nearly collided with a stopped train and was almost struck from behind by a third train. The track circuit that malfunctioned was identified. At that time a computer program was developed to scan for circuit faults. The system was scanned once a week. After not finding anything for a year, the computer scan was dropped to once a month. After the accident in 2009, the computer check was increased to twice a day. Track circuits also failed to detect trains (without incident) in March and May 2009. And faulty track circuits caused 337 delays for the 12-month period ending June 2008. In 1999, the entire Washington Metro system operated in manual mode for 20 months when 20,000 track block relays were replaced after a handful failed.

A long-term goal of Positive Train Control is complete replacement of all track circuits, signals, and relays (like those used on Washington's Metro) with GPS and radio control. Removing trackside signals will reduce maintenance costs and hopefully improve reliability. Track circuits, however, will not disappear any time soon.[10]

Stopping a Train

Braking is simply not a repeatable process. Microscopic examination of the steel surfaces involved shows sharp peaks and valleys. Every time a train moves over the track the surfaces are microscopically polished into different surfaces. As the wheel wears, the contact pattern can change from a single location to two (Figure 6.1). Wheel-rubbing also varies on curves with different radii, on misaligned track, and on loose track that deflects under load. Braking also depends on rail surface contaminants such as water, ice, leaves, and so forth.

The maximum braking force available (other than emergency braking) is called a full service brake application. There is only so much air pressure available in the system. Once it is used up, the brakes have been maxed out (see Chapter 10). Full service braking varies from train to train, depending on circumstances and the available air pressure. Air pres-

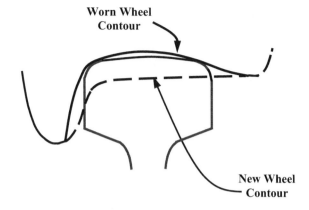

FIGURE 6.1. Worn wheels change contact from a single spot to two.

sure will be lower if the train has just applied its brakes. Also, the pressure in the system is highest up front near the air compressors supplying the pressurized air; because of leakage, it is lowest in the back of the train. Leakage depends on the length of the train, the condition of the seals, and the ambient temperature.[11] All of the above factors affect how fast the air pressure signal travels the length of a 1- to 2-mile (1.6- to 3.2 km-)-long freight train.

Trains are permitted to temporarily operate with 15% of their brakes not operating. Depending on the brake rigging and brake cylinder, a brake shoe can be "sticky" or even stuck open. These same effects alter the clamping forces on each brake shoe and the actual braking forces.

FRA COMPUTER SIMULATION predicted that a grain train with 123 loaded cars 6,798 feet (2,072 m) long and weighing more than 14,000 tons (12,700 MT) must begin braking 1,762 feet (537 m) sooner than a train driven by a skilled locomotive engineer. Earlier braking is required to meet the required PTC reliability of 99.9995% for not overshooting. The computer concluded that the same grain train traveling at 30–60 mph (48–97 km/h) will stop more than 1,000 feet (305 m) from the target 62% of the time, assuming the train's weight is known with 99.9999% reliability.

Stopping 1,000 feet too soon in the middle of nowhere is not a problem.

Stopping too soon in the industrial heartland can block other tracks, adding to congestion and accident potential. PTC with conservative braking algorithms is actually expected to slow down the rail network.

Braking Algorithms

Research to improve the braking problem is ongoing. Instead of one computer algorithm that conservatively stops all trains, the computer program is written to customize itself for each individual train.

The braking algorithm uses sensors to measure air pressure and braking decelerations. After the brakes are applied on a train for the first time on each trip, the algorithm recalibrates itself based on the measured performance. It is hoped the inaccuracy of braking can be reduced by 50%. Braking variations will still occur on curves, when the rail's surface changes, and with track misalignments. Electronically Controlled Pneumatic (ECP) brakes, described in Chapter 10, will also reduce braking variations.

Locating the Train

The Global Positioning System (GPS) is a military system using radio signals transmitted from satellites to identify location. The GPS receiver precisely measures the signal transit time from four to six satellites to calculate position. The system originally sent an accurate signal reserved for the military and a degraded signal with 100-m (328-foot) accuracy to everyone else.

Because of increasing civilian use, the U.S. Coast Guard developed a system that broadcasts a land-based differential correction signal. This new system was known as differential GPS, or DGPS. The accuracy improved to 1 m (3 feet) at the base of a DGPS transmitter and degraded an additional 1 m (3 feet) for every 150 km (93 miles) from the DGPS tower. As the number of DGPS transmitters increased, the system became known as the National DGPS, or NDGPS.

If a NDGPS transmitter goes down, the system automatically selects a backup transmitter—a configuration known as dual station coverage. There are currently not enough transmitters to blanket the entire country with dual station coverage.

NDGPS is considered by the FRA to have an accuracy of 10–16 feet

(3–5 m) with 95% confidence, not accurate enough to identify which of two parallel tracks a train is on. High Accuracy NDGPS is under development with a stated resolution of 3.5–12 inches (10–30 cm).

GPS signals are lost in tunnels and in urban canyons. Reliability decreases after an accidental system shutdown. The required reliability for PTC locating trains is 99.9999%.[12] To raise the reliability of NDGPS from 95% to more than 99.9999%, more information needs to be processed. Wheel rotations are counted with a tachometer. The wheels do not perfectly roll but actually slide a small amount that varies with friction and the shape of the worn surfaces. A tachometer accumulates errors but is sufficiently accurate for a short distance. The tachometer is supplemented with readings from an accelerometer that records all changes in speed to help keep track of the train's location. (Accelerometers are explained in Chapter 11.)

A computer can distinguish between two or more parallel tracks if it keeps track of every time the train changes tracks. This is done with an onboard gyroscope.[13] The PTC computer contains a database of the track layout, including the radiuses of all curves and turnoffs. Data from the tachometer, accelerometer, gyroscope, GPS, and computer map are now claimed to reliably locate trains within 3 inches (7.6 cm).

Obviously, a 2-mile (3.2-km)-long train needs more space than a 2,000-foot (609-m)-long train. Railroads already have the length of each train computerized to calculate required horsepower, braking, and billing. One might think customer-billing is reason enough to keep track of the train's length. Not so, say the engineers. Existing systems are not reliable enough (i.e., not 99.9999%). New sensors are being developed to solve this problem.

COMMUNICATIONS SYSTEMS

Historically, railroads have operated complex and enormous communications systems ranging over many hundreds of miles. After waves of bankruptcies and mergers, the remaining major companies must now communicate along tens of thousands of miles of track.

Trainmen talk to each other while switching cars in yards. The dispatcher authorizes routes and coordinates train meets, maintenance work, and any delays. Repair crews move around equipment and supplies.

Railroad voice communications are transmitted by radios, telephone wires, and fiber optics.

Railroads also transmit signals to control train movements. Signal lights stop, slow, speed up, or warn trains of future signals. Other signals remotely control track switches. As explained in Chapter 5, the signal logic that prevents train collisions can be quite complex. Train signals are transmitted by the same means used for voice communications. Signals can also travel as electrical pulses through the rails.

Trackside sensors detect hot bearings, out-of-round wheels, dragging equipment, hot wheels, high water, deep snow, and rock slides, and transmit a warning to the train crew or dispatcher. Today trains use radio signals to activate the emergency brakes in the back of the train and to remotely control additional locomotives on long trains.

Railroad communications is a massive hodgepodge of different needs and technologies of different vintages. Currently, railroads operate about 16,000 base transmitters, 90,000 mobile radios, 125,000 hand-held radios, 12,000 trackside defect detectors, 100,000 signal lights, perhaps 50,000 signaled road crossings, and thousands of remotely controlled switches.

The Radio Spectrum Is Crowded

All electromagnetic radiation (X-rays, thermal radiation, gamma rays, visible light, radio waves, etc.) travels at the speed of light with variable frequencies or oscillations per second known as Hertz.

The Federal Communications Commission (FCC) regulates the radio spectrum in the United States. Fundamentally, its purpose is to protect the integrity of vital government communications such as the military, air traffic control, police, and firemen.

A license from the FCC allows transmission at a specific frequency (i.e., 120 megahertz, or MHz, or 120 million cycles per second). Each transmission channel is separated by a 25-kilohertz bandwidth. The FCC will approve two transmitters operating at the same frequency if they do not interfere. Interference depends on distance between transmitters, transmission power, how close the two transmitted frequencies are, and local topography. Avoiding interference also depends on the details of antenna design and many other engineering issues.

A pair of two-way radios operates at two frequencies so that both radios can transmit simultaneously, similar to how two people normally talk. Almost all railroad radio communication requires two-way transmissions and therefore requires a pair of licensed frequencies.

Up until the late 1980s, there was more spectrum than demand and radio spectrum was free. Beginning in 1994, the FCC began auctioning frequencies. With the explosion of cell phones and other wireless devices, the problem is getting worse.

In the late 1980s, hundreds of individual licenses were granted to the railroads for thousands of base stations transmitting 6 pairs of frequencies in the 900-MHz band.[14] These frequencies were reserved in anticipation of future PTC radio transmissions. In 2000, the American Association of Railroads (AAR) petitioned the FCC to combine all the licenses into one single license to be administered by the AAR. The rationale was that debugging new PTC systems would require frequent relocation of many new transmitters. The railroads were given control of a 70-mile (112-km) strip on both sides of the track (140 miles [225 km] wide) along all main routes on the national rail network.

In 2001, a 120-mile (193-km) segment of track in Illinois was targeted for development for 110-mph (177-km/h) passenger service. The plan was to design a new PTC system that operated at 900 MHz.

Positive Train Control requires the transmission of significantly more information. The PTC data link must constantly update the train's location and speed; calculate the ability to stop; communicate with track side signals, switches, and central computers; process all data; and transmit continuous permission to continue advancing. If new authority is not granted in a timely manner, the train will default to a failsafe condition and stop, as will all following trains on the same track.

If a vital system component fails, the computer must activate backup systems. This adds to the communications logjam by requiring all components to constantly verify they are working properly, the so-called heartbeat signal. Many systems use a 20-second failsafe activation; in other words, the signal retransmits 3 times, taking up to 20 seconds before alternate plan B (or C or D) is activated.

The Illinois PTC system did not meet its objectives. A large part of the problem was communication software development. The 900-MHz data link system could not maintain a reliable flow of information. The Illinois PTC project was shut down at the end of 2006 when Illinois withdrew its support.

A very similar story occurred on an Amtrak route in Michigan. After several years of development a PTC system became operational in 2000. Speeds were raised to 95 mph (153 km/h) in 2002. Plans to operate at 110 mph (177 km/h) were repeatedly announced and repeatedly delayed until finally occurring in early 2012.

Meanwhile, cell phone use exploded and gobbled up unused 900-MHz spectrum. The railroads had to start over. In October 2008, four major railroad companies agreed to standardize PTC at 220 MHz. The problem is especially acute in metropolitan areas, where the radio spectrum is most crowded. Railroads are still hoping for FCC intervention to obtain more spectrum.

To create more channels the FCC is phasing in a reduction of the current channel spacing of 25 kHz to 12.5 kHz by 2013. All existing radio transmitters must be replaced to transmit on 12.5 kHz channels. A further reduction to 6.25 kHz is anticipated in the future.

Meanwhile, the railroads are scrambling to obtain required spectrum. Los Angeles's Metrolink, working with Amtrak, Union Pacific, and BNSF, agreed to fast-track PTC in metro Los Angeles by 2012. (Recall the 2008 collision in Los Angeles that killed 25; see the beginning of Chapter 1.) As of early 2012, Metrolink was still battling a year-long legal challenge blocking the FCC from assigning 40 channels of 220 MHz spectrum for its new PTC system. The Southern California Regional Rail Authority is struggling to design a resilient communication network with sufficient redundancy to be 99.99% available even after cable cuts, earthquake, fire, and power failures. With key technology still unavailable, the 2012 deadline (promised since 2008) has now officially slipped into 2013.

To reduce wireless PTC transmissions on crowded spectrum new methods are being used to transmit information. In Los Angeles a communication network is being designed with digital microwaves, fiber optics, and copper cables. New software is required to manage the digital system and provide system redundancy.

Analog versus Digital

The world is an analog place with infinite resolution. The computer world is digital; everything is described with a binary system— 1's (on) or 0's (off). Computer resolution is limited to the maximum number being processed by the computer. An old-fashioned analog image on film can be magnified to the limits of a light microscope, whereas magnification of a digital image is rapidly limited by the number of pixels available.

An analog signal is digitized by sampling the signal and converting the analog signal into numbers suitable for computer input. The more numbers sampled, the better the digital representation of the analog signal, but of course it takes more computer effort to process more information. As the computers became bigger and faster, digitization became increasingly more accurate and useful.

Why bother to digitize analog information? Once the analog information is inside the computer, it can be manipulated. In the case of digital images, red eye can be removed or blurry images can be sharpened.

The opportunities for the railroads are enormous. Once all the signals are digitized and collected in one place, the entire rail network can be optimized. The communication problems for the railroads are immense, however, and Positive Train Control remains to be conquered.

Currently, all railroad transmissions are designed and optimized for voice analog communications. All signal, data, and voice communications must be digitized. The railroads must meet the FCC deadline to convert all radio transmitters to 12.5-kHz narrow band by 2013, design for the 2015 FRA PTC deadline, and anticipate future bandwidth reduction to 6.25 kHz.

Although PTC data transmission is becoming more important, railroad voice communications will never disappear. New digital radios will handle both voice and computer data and operate within a 6.25-kHz bandwidth. Standard design problems (coverage, signal propagation, and interference) must be resolved for this new digital radio equipment. Railroads also require added reliability for this safety-critical application and terrorist-proof security. An estimated 100,000 new digital radios are required to make 60,000 miles (96,500 km) of track operate with PTC.

Vital Systems

The concept of vital failsafe systems is well established in the railroad industry. For example, the electro-mechanical relay described earlier (see Figure 5.3 in Chapter 5) uses gravity to activate the stop signal. If any failure disrupts the flow of electricity, the system automatically defaults to a failsafe condition and signals "stop."

A vital PTC failsafe system must have a backup plan for all anticipated failures. A computer-controlled vital system must automatically transfer control to a backup computer. Unplug both computers and the train must safely stop itself.

Positive Train Control consists of hardware, software, and communications (i.e., more hardware and software). A vital PTC system must be designed to safely operate with random hardware and software failures. To accomplish this all critical functions must be performed in a variety of ways using multiple software and hardware for system redundancy.

Two identical hardware units may run the same software and compare results, or two software programs, developed by independent teams, may be used for software redundancy. All software must reboot or re-initialize itself and pick up where it left off after random power failures. As a general rule, complex software does not work reliably until extensively debugged in real use.

Railroads can use a non-vital system that sits on top of their existing train control system. If the railroad currently uses track warrants to authorize train movements, the PTC is added on top as an "overlay." The dispatcher still authorizes train movement by radio to the train crew. The PTC system intervenes if the crew misses a command to stop. In this case a simpler, less redundant PTC system can be used. If the PTC system fails, the existing control system is still in place as a backup. A complex failsafe "vital" PTC is not required. The FRA requires the non-vital system to demonstrate with analysis an 80% reduction in accidents.

Roger W. Baugher in *Railway Age* explains the difference between vital and non-vital with a highway analogy. If car traffic is controlled at an intersection with a traffic light and the traffic light fails, everyone understands that the intersection is now a four-way stop. Traffic continues, albeit less efficiently. The traffic light is a non-vital overlay. If, on the other hand, cars are spaced on a highway every 100 feet (30 m) at 70 mph (113 km/h) with computer control, the control system must be a safety critical vital system. If anything breaks, all traffic must stop. The non-vital system is less complicated and less expensive; the vital system is more efficient.

Different railroads are responding differently to the choices of vital, non-vital, overlay, or stand alone. It may depend on how happy they are with their current signal system or on anticipated future growth. These many choices contributed to another problem holding back PTC.

INTEROPERABILITY

Main track is typically owned and operated by one railroad company, but is usually shared with many railroad companies. Computer and communications must be standardized to allow one company's loco-motives to pass seamlessly into another company's territory. All railroad companies, the FRA, and equipment suppliers have to agree on technical standards for shared operations.

When Congress passed the Rail Safety Improvement Act in 2008, there were nine different railroads operating incompatible PTC systems under various stages of testing and development. Besides vital or non-vital considerations, railroad companies approached the communication and control problems with a variety of schemes. Are communications and controls located mainly on the locomotive, along the track, or at a central office?

After three years of study and negotiations, the AAR established rules for interoperability at the end of 2008; it essentially agreed on what must be standardized. Because every railroad was approaching the PTC problem with different communication and computer architecture, the problems of interoperability proved more difficult than expected. Two years later, only 3 of 40 detailed engineering specifications had been finalized. (Specifications are required for communication protocols, data management, messaging, displays, interfaces, etc.) Without completed specifications, prototype 220-Mhz digital radios are not even expected until 2012. System testing has started, with temporary radios operating at the wrong frequencies. Recall that 99.9999% reliability is required. And compare that to the reliability of today's cell phones.

A similar problem of interoperability exists in Europe. The current European plan calls for establishing an interoperable PTC system on just 15,000 miles (24,000 km) of track (compared to 60,000 [96,500 km] in the United States) by 2020.

In March 2011, the president of the American Public Transportation Association (APTA) released a statement: "an overwhelming majority of the nation's commuter railroads" have concluded that technology issues, radio frequency availability, and lack of federal funding "make nationwide implementation by the deadline impossible."[15] APTA recommends a three-year delay in implementing PTC.

In May 2011, the Federal Communications Commission (FCC) requested comments on PTC problems. The FCC declared "difficulties in meeting the spectrum needs for PTC are directly related to the geographic areas where PTC must be deployed and the current spectrum environment in those areas." The FCC further acknowledges that "PTC system[s] . . . may differ depending upon the frequencies and amount of spectrum available, so we seek information regarding how the use of different frequencies and amounts of spectrum could affect . . . a PTC system. We also seek input regarding how . . . rail operations affect the amount of spectrum . . . needed to successfully implement PTC. Such factors could include speed . . . , the number of track lines in a certain geographic area, . . . rail traffic . . . , separation of trains, and the hours of operation . . . The Bureau also seeks comment on the spectrum coverage area that will be needed by the railroad industry for PTC implementation."[16] In other words, the FCC agrees that it is a big problem.

ASSESSING RISK

The FRA requires a risk analysis for all new PTC systems. Each railroad must document expected reliability of the new PTC equipment based on engineering analysis and testing and estimate the probability of accidents with the new system.

One way to study risk is to identify the number of opportunities for accident scenarios. Accident scenarios are the number of times trains approach, pass each other, or travel too fast on curves.

Computer software is used to simulate a company's entire rail network.

The software inputs are track layout and train schedules. The simulated dispatcher routes trains through the network by authorizing movements and organizing train meets. Trains with higher priority are allowed to pass lower-priority trains. The model even includes random delays. Trains accelerate and decelerate according to $f = m \times a$. Trains advance with small changes in speed until all train forces, including gravity, wheels rubbing on curves, traction, and braking forces, are balanced.

The model output is the number of potential accident opportunities for an entire rail network. The probability of an accident occurring is the number of accident scenarios multiplied by the probability of the humans and PTC system both failing.

The risk analysis also requires prediction of failure rates for individual PTC components, subsystems, and software. Each component must be tested individually and as part of a system. A computer simulation of the locomotive (with input throttle and braking) is used to interact with other parts of the system during "laboratory integration testing." Since the humans will interact with the new technology in different ways, all potential human errors have to be considered and studied as well.

After completion of lab testing, the system is tested in the real world with controlled train operations. Speeds are carefully monitored, and the wireless connection is tested at many locations by issuing artificial train orders. The FRA must approve the next step—revenue service testing, which involves a fully operational system but still under intense scrutiny. Past experience has shown that this phase takes 6 to 12 months to debug all problems and gain confidence in the new system. Needless to say, designing and implementing a PTC system is a lengthy and expensive process.

PTC AND THE FUTURE

Efficiency and throughput can be improved if trains are spaced closer.[17] Currently, to avoid collisions trains are spaced by track segments, or blocks. The block's length is based on the distance it takes for the most difficult train (usually the train with the highest kinetic energy) to stop safely. With PTC each train can have a movable customized block—a protected zone—attached to it. Only the most difficult trains will require the longest block. Most blocks can be shorter; thus, trains are spaced closer together and capacity is increased. The next generation of PTC is expected to include movable blocks.

Debugging all of this new hardware and software will be difficult but not necessarily unsafe. Recall that everything is designed to fail safe. Unfortunately for rail network operations, the safest condition for a train is "stopped." If anything fails—software, hardware, or the communications data link—the train will stop. Trains stopping just a little bit more often will quickly make the system less efficient. Operators are also expected to travel a few mph slower than normal to avoid computer intervention. Furthermore, recall the difficulty in stopping a train accurately. If the train stops prematurely, it can block other tracks, creating gridlock. The near-term expectation is for PTC to decrease system efficiency.

One might think computer control is close to replacing the train crew. PTC will not, however, respond to something on the tracks. The crew is also needed to nudge the train forward when PTC stops the train too soon. There are many other duties for the train crew. If the train stops on a hill, hand brakes must be applied to prevent a runaway. The crew is trained to make minor repairs (i.e., broken air hose or coupling) and interact with emergency responders during a derailment. The crew must also examine the train on every curve for shifted loads, dragging equipment, sticky brakes, hot bearings, and so forth. For any equipment failure (i.e., overheated bearing), two people are needed to break the train apart and set out the bad car.

There are no plans to replace the crew. The train crew will still drive the train and must back up and debug the new PTC systems. The thinking is that humans cannot be expected to quickly respond in an emergency if they have nothing to do 99% of the time. Also, any suggestion of replacing or reducing train crew will create unwanted labor problems that have historically taken many years to resolve. Starting up this new and complex equipment with a disgruntled workforce is unthinkable.

7

Moving at the Wrong Speed

B elieve it or not, it's possible to derail a train by going too slow—more about that later.

TOO FAST ON A CURVE

In 1947, a Pennsylvania Railroad passenger train with 2 steam locomotives and 14 cars left Pittsburgh at 1:05 a.m. bound for New York City. The train had just descended a steep 1.73% grade when it overturned on a sharp 8.5-degree curve (675-foot [205-m] radius) at 3:20 a.m.[1] The speed limit downhill was 35 mph (56 km/h) and 30 mph (48 km/h) on the curve. Instructions required the train crew to test their brakes 2 miles (3.2 km) before the curve.

The 2 locomotives plunged down a 92-foot (28-m) embankment with 5 cars attached. Ten of the 14 cars derailed. Twenty-four people were killed. The investigators concluded that excess speed caused the train to overturn on the curve. The overturning speed was calculated to be 65 mph (105 km/h). Elsewhere in the news on the same day as the accident, the Pennsylvania Railroad, the largest railroad in America, reported operating losses for the year 1946—their first ever.

Speeding trains overturning on a curve also occurred in California in 1956 (killing 30) and in Virginia in 1978 (killing 6).

INERTIAL LOADING

Everyone knows, or thinks they know, what centrifugal force is. It's the phenomenon that flings passengers against the car door on a curve, the force that keeps the water in the bucket when swung fast enough overhead, and the force that derails trains on a curve. But centrifugal force can be a source of much confusion because it's not a force in the traditional sense. Centrifugal force is an inertial effect that occurs when a body in motion changes direction, as in each of the examples above.

Per Isaac Newton, a body in motion tends to stay in motion. If somehow we could eliminate gravity and air resistance, a ball thrown straight up would continue straight up forever. It takes additional force to change the straight-line motion of the ball and to move a train around a curve.

Inertia, the property of matter that resists changes in motion, is most easily explained by accelerating in an elevator. If a 100-lb (0.44-kN) person is standing on a scale in an elevator accelerating up, the scale reads something higher than 100 lbs. If the elevator is accelerating down, the scale reads something less than 100 lbs. If the elevator is accelerating up at 16 ft/sec^2, or one-half the normal acceleration of gravity, the scale will read 150 lbs (0.66 kN). The extra 50 lbs (0.22 kN) is from the person's body resisting acceleration.[2]

When a body accelerates, or changes velocity, that acceleration is accompanied by a force. According to Newton's Second Law, $f = m \times a$. The body's inertia ($m \times a$) is not a force even though it acts on the scale like a force. The additional 50-lb reading on the scale is the 100-lb person's resistance to accelerating up 16 ft/sec^2 (4.9 m/s^2)—the person's inertia.

Inertia always acts in the opposite direction of the acceleration. In the case of the elevator, the person is accelerating up and the inertial response is acting down and is being recorded by the scale. A similar thing happens in circular motion. Circular motion at constant speed creates an acceleration that points toward the center of rotation.

We tend to think of acceleration as being a change in speed (see Chapter 4). Velocity is actually a vector with both direction and magnitude. (The velocity vector's magnitude is also known as speed.) Any change in

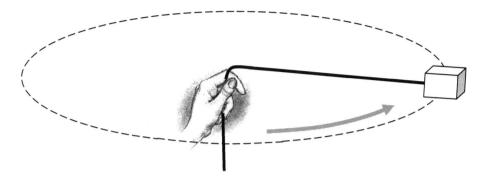

FIGURE 7.1. A string exerts a centripetal force on the block rotating on the end of the string.

the velocity vector, be it a change in speed or a change in direction, requires a force to create the change.

Consider a ball rolling along a straight line. One could constantly tap the ball with a stick, forcing it to move in a circular path. The tapping force, always pointing to the center, is changing the ball's velocity vector's direction. The ball is moving at a constant speed but changing direction; the ball is said to be accelerating toward the center of the circle.

Consider a 1-lb (0.45-kg) block rotating on the end of a 4-foot (1.2-m) string in a horizontal plane at a constant speed of 20 ft/sec (6 m/s). The direction of the velocity vector, always perpendicular to the string, is constantly changing and creating acceleration toward the center of rotation (Figure 7.1).

Acceleration for circular motion equals the velocity squared divided by the radius of the circle, or 100 ft/sec² (30.5 m/s²). Per Newton's Law, the string must exert a force on the block equal to m × a, or a force of 3.1 lbs (13.8 N) toward the center of rotation. (Recall that to correctly calculate f = m × a the weight must be converted into a mass by dividing by the acceleration of gravity—32.2 ft/sec² [9.8 m/s²].) The force the string exerts on the block is called the centripetal, or center-seeking, force. The block exerts an inertial load on the string, keeping it tight.

The so-called centrifugal force is not a force; it's the block's inertial resistance to the centripetal acceleration. The 1-lb (0.45-kg) block resists the acceleration imposed by the string just as the person in the elevator

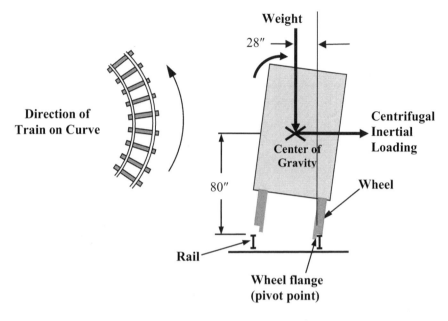

FIGURE 7.2. The locomotive shifts to the outside of the curve and engages the outer wheel flange, as shown. Upon overturning, the right wheel acts as a pivot and the left wheel lifts.

resists the upward acceleration. The term *centrifugal force* is incorrect. We will use the term *centrifugal inertial loading*. But, of course, the so-called centrifugal force feels like a force when holding on to the string attached to the rotating block.

A locomotive moving around a curve is similar to a rotating block on the end of a string. Both experience acceleration toward the center of rotation. The inertial loading keeps the string tight and creates a lateral force on the locomotive at the wheels. Lateral forces between the wheels and the rail must react against the centrifugal inertial loading to keep the train on the tracks.

If the centrifugal inertial loading is excessive, the locomotive begins to tip. The flange of the wheel catches on the rail and the locomotive starts to rotate, as shown in Figure 7.2. In fact, that's why the flanges are on the inside of the wheels. If the flanges were on the outside, the slightest bit of wheel lift would slide the locomotive off the tracks.

In the 1947 Pennsylvania Railroad overturning accident, the locomotive weighed 320,000 lbs (145,150 kg). The centrifugal inertial loading

of the locomotive moving at 88 ft/sec (60 mph [97km/h]) on a curve
with a radius of 675 feet (206 m) is:

$$\frac{320{,}000 \text{ lbs sec}^2}{32 \text{ ft}} \left(\frac{88 \text{ ft}}{\text{sec}}\right)^2 \frac{1}{675 \text{ ft}} = 114{,}000 \text{ lbs } (507 \text{ kN})$$

The centrifugal inertial loading is trying to tip the locomotive clockwise about the pivot point (the bottom of the right wheel). This rotation
is resisted by the weight of the locomotive (also acting through its center
of gravity), which tries to rotate the locomotive counterclockwise.

The locomotive's weight and inertial load both exert a torque. A torque
is a twisting force applied to the end of a lever arm that tries to tighten a
nut. A 10-lb (44.5-N) force on the end of a 9-inch (23-cm)-long wrench
exerts a torque of 10 × 9 = 90 inch lbs of torque (10 Nm).

The Pennsylvania locomotive had a center of gravity 80 inches (2 m)
above the rail. The centrifugal inertial loading tries to rotate the locomotive with a clockwise torque equal to 114,000 lbs × 80 inches—more than
9 million inch lbs of torque (6.3×10^6 Nm).

The lever arm for the locomotive's weight is halfway between the rails,
or 28 inches (0.7 m). The torque from the locomotive's weight that tries
to resist the overturning torque from the centrifugal inertial loading
equals 320,000 lbs × 28 inches—almost 9 million inch lbs of torque.

The torque trying to overturn the locomotive is slightly larger than the
torque from the locomotive's weight resisting the overturning torque.
The locomotive is just starting to overturn at 60 mph (97 km/h).

SUPERELEVATION

The outside rail on a curve is usually higher than the inside
rail. The elevation of the outside rail relative to the inside rail is called
superelevation.

A raised outside rail rotates a locomotive counterclockwise and helps
fight off the clockwise rotation from the centrifugal inertial loading, at
least a little bit. In fact, if the car is made top heavy and the right wheel
is lifted enough (even at zero mph), eventually the car tips over counterclockwise. The car tips over at zero mph when the weight load points
outside the inner rail, as shown in Figure 7.3.

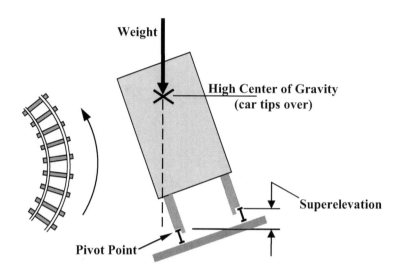

FIGURE 7.3. A car will tip over toward the inner rail, even at zero mph, if the superelevation and/or center of gravity are raised enough.

In 1947, the investigators concluded that the locomotive would overturn on the curve (with outer rail raised or superelevated 3.5 inches [8.9 cm]) at 65 mph (105 km/h).

Amtrak's 150-mph (241-km/h) Acela creates its own bank angle by tilting up to 4.2 degrees. If the Acela is operating on a curve whose outside rail is raised 2 inches (5 cm), the Acela can speed as if it is on a curve that is raised an additional 7 inches (17.8 cm) higher—for a total superelevation of 9 inches (22.9 cm).

Tilting trains are far more complicated and not the first choice of railroad companies. It is easier to operate on redesigned curves with larger radiuses. Of course, curves with larger radiuses take up more real estate—difficult to do in older, built-up neighborhoods.

TOO FAST ON A TURNOUT

Far more common is moving too fast on a turnout. On a turnout the track crosses over with sharp turns to merge onto a parallel track. The engineer must slow the train for the turnout or risk overturning. Just such an accident happened in 1951 in New Jersey, killing 84.

Construction of the New Jersey Turnpike required relocating the train tracks 60 feet (18 m) north for a few months. The temporary track was about 2,800 feet (853 m) long and contained a 57-foot (17.4-m) temporary wooden trestle anchored on both ends by massive concrete abutments. The trestle was also part of the turnout, a 121-foot (36.9-m)-long curve with a radius of about 1,100 feet (335 m).

The speed limit on the main track was 65 mph (105 km/h). The temporary track went into operation for the first time at one o'clock p.m. on the day of the accident, February 6, 1951. The speed limit on the turnouts and temporary track was 25 mph (40 km/h).

The rush hour train with 11 cars was particularly crowded with about 1,000 passengers, many standing. The locomotive and first seven cars derailed. The third and fourth cars were the most damaged. Those two cars struck the concrete abutment (knocking off a big hunk) and fell down the 25-foot (7.6-m) embankment. The third car crashed onto its side, its center sill broke, and the roof and both sides were badly damaged. The right side of the fourth car was torn open its entire length. The investigators concluded that the locomotive's speed exceeded the calculated 76 mph (122 km/h) overturning speed.

Too fast on a curve is by no means an obsolete problem. In a nearly identical accident in Chicago on September 17, 2005, a commuter train went off the tracks at a turnout, killing two. The engineer missed the signal to slow from 70 to 10 mph (113 to 16 km/h).

Too fast on a curve should be prevented in the future by Positive Train Control (see Chapter 6).

DERAILING ON CURVES

Before reaching the overturning speed, a slow, heavy freight train is far more likely to derail on a curve by rail rollover, wide gage, or wheel climb (Figure 7.4).

The tracks are constantly moving around (and constantly being readjusted) because of settlement and train forces (Figure 7.5). The track spikes do not prevent the rails from overturning but do keep them from spreading. Rail overturning is thwarted by the downward wheel forces. If the wooden cross ties are rotted, inertial loading on curves may widen the rails.

Standards are established for maximum distance between rails (gage),

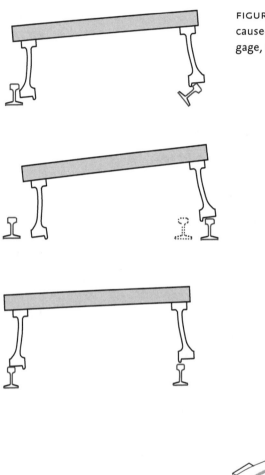

FIGURE 7.4. Derailments can be caused by rail roll over, wide gage, or wheel climb.

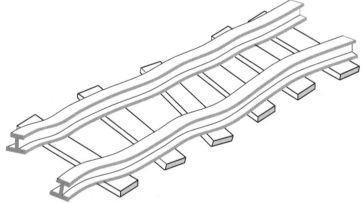

FIGURE 7.5. The rails and cross ties can rotate, warp, and shift (exaggerated).

maximum dips in each rail (profile), and maximum deviation from straightness (alignment). Higher classes of track require tighter requirements to operate safely at higher speeds. For example, freight trains are limited to 40 mph (64 km/h) on Class 3 track and 60 mph (97 km/h) on Class 4 track. (The track classes are reviewed in Chapter 11.)

Although track geometry today is measured automatically with high-speed cars using laser sensors, the standards are based on low-tech methods of measuring the deviation from a 62-foot (18.8-m) string pulled tight. Every 62 feet (18.8 m) of Class 3 track can deviate up to 1.5 inches (3.8 cm) from straight and dip up to 2.25 inches (5.7 cm). The Acela operates at 150 mph (241 km/h) on Class 8 track. Every 31 feet (9.4 m) of Class 8 track can deviate up to 0.5 inch (1.27 cm) from straight and dip up to 1 inch (2.54 cm).

Class 8 track geometry is checked every 30 days. In fact, when Amtrak was preparing to operate Acela at 150 mph, Amtrak's chief engineer of maintenance, the director of track geometry, and many others rode the geometry car every two weeks for months. They considered it a bonding experience.

The operators will also report any rough or shifted track as it occurs. For all trains operating above 125 mph (201 km/h), at least one train per day has sensors to measure, quantify, and record the location of any rough track.

Concrete, instead of wood, is used for ties on Class 8 track. The concrete is less susceptible to shifting and water damage. At least once annually Class 8 track gage stability is checked with a special car that loads the rail sideways with a force of 10,000 lbs (44.5 kN). Class 8 track is also inspected twice a year with ultrasonic sensors for internal fatigue cracks.

L/V RATIOS

The tendency to derail is often described by the L/V ratio, where L is the lateral force and V is the vertical force at the wheel-track interface, as shown in Figure 7.6. The higher the L/V ratio, the more likely the car is to derail.

There are rough guidelines for L/V limits. Wheel climb may occur if:

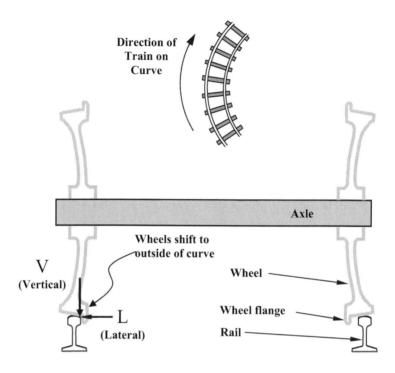

FIGURE 7.6. Illustration of the ratio of lateral to vertical forces at the wheel-rail interface. Notice the wheels shifting to the left when going around the curve.

L/V greater than 1 for new freight cars with new wheels on new, straight track

L/V greater than 0.82 may be unstable on curves

L/V greater than 0.75 can be unstable for worn wheels and worn rail

L/V greater than 0.68 may overturn a poorly constrained rail

Rails spaced too close together can also encourage wheel climb.

The stated L/V ratios are merely rules of thumb, not rigid predictors. There are many other factors that interact, such as the condition of the trucks, rails, and wheels and whether or not the car body is bouncing on its suspension.

The L/V ratio can also vary greatly as the wheels and rail wear and as the contact location changes. A worn rail on the outside of a curve is

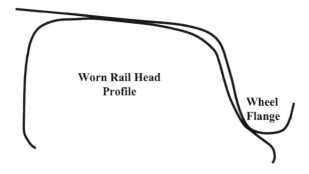

FIGURE 7.7. A worn rail head shifts contact and the L/V ratio.

shown in Figure 7.7. Another wear pattern is shown in Figure 6.1 in Chapter 6.

Acela Class 8 track must be checked annually with an instrumented car that measures the L/V ratios.[3] Special load sensors are on the truck frame and on the floor of the car. If the L/V ratio exceeds 0.6, the speed must be reduced until repairs are made.

HANDLING LONG, HEAVY FREIGHT TRAINS
The Ends Travel at Different Speeds

One might think the back of the train ought to know what the front of the train is doing, but not so for long freight trains. The couplings between each pair of cars can move plus or minus 6 inches (15.2 cm) from their neutral position for a total movement of 1 foot (0.3 m). A train with 100 cars can run in or out 50 feet (15.2 m), for a total change in length of 100 feet (30.4 m). This 100-foot movement is known as slack. When the slack runs in (or out), different parts of the train move at different speeds. This creates additional in-train forces that can break a coupling or derail the train.

Where does the slack action come from? Shortly after inventing the automatic airbrake (see Chapter 10), George Westinghouse invented the friction draft gear to damp the shock of stopping long trains. The airbrake signal activated the brakes in the front of the train first. Cars in the front stopped, but the cars in the back continued moving.

Draft gears use friction wedges and springs, as shown schematically in

BOTTOM VIEW (Lying on rails looking up)

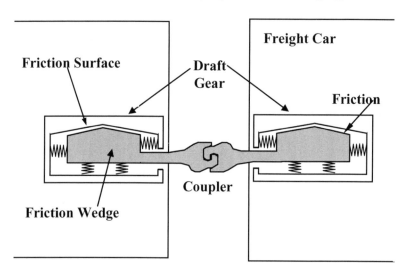

FIGURE 7.8. Bottom view of the schematic representation of friction wedges and springs in draft gear. The actual configuration is slightly different; see note 4.

Figure 7.8. The vertical springs on the bottom press the friction wedge against a friction surface, creating forces as the draft gear moves in and out.[4] The horizontal springs resist the back-and-forth movement and return the draft gear to a neutral position when the load is removed. A coupling is shown in Figure 7.9.

For more delicate cargo additional cushioning is provided with hydraulic end-of-car cushioning devices. The hydraulic cylinders, replacing the draft gear, can travel 10 to 18 inches (25.4 to 45.7 cm), depending on the actual device. As the number of cars increases, the slack becomes excessive and very difficult to control. For this reason, the number of cars with hydraulic cylinders on a train is limited.

The slack serves or has served many purposes. Before the widespread use of roller bearings on freight trains, slack was required to start a long, heavy train. The older-style plain journal-bearings started with metal-to-metal rubbing and tremendous starting friction (see Chapter 8). The engineer would first compress or bunch the train, and then, pulling in the opposite direction, each car would start up one at a time in serial

FIGURE 7.9. Coupler, coupler shank, and striker. The draft gear is not visible.

fashion. The starting friction in each car would be overcome one car at a time instead of all at once.

Originally, cars were connected with a pin-and-link connection. To connect cars a trainman had to stand between the cars and insert a pin. (Doing so was as dangerous as it sounds. Over the years thousands have died coupling cars.) The whole point of the modern coupler (circa 1898) is to couple cars without a person standing between the cars.

To couple cars with modern couplers one car is shoved into another. With the brakeman no longer standing between the cars during coupling, the locomotive crew became less careful coupling cars. The draft gear was needed to protect train equipment and cargo from coupling impacts.

The suggested maximum coupling speed is 4 mph (6.4 km/h). Speeds

of 8 mph (12.8 km/h) or higher, however, are not unheard of. The force
to compress a draft gear into metal-to-metal contact varies from 200,000
to 300,000 lbs (889 to 1,334 kN), depending on the model. Test data on
a loaded tank car weighing 286,000 lbs (129,700 kg), the maximum allowed car weight, coupled at 7 mph (11 km/h), created coupling forces of
560,000 to 960,000 lbs (2,490 to 4,270 kN), depending on the draft
gear's age. Older, softer draft gears go solid sooner, creating higher-im-
pact forces.

Draft gears are also rated for their ability to dissipate energy. The definition of work is force × displacement (see Chapter 9). An example energy rating for a draft gear is 36,000 foot lbs (48.8 kJ). Each draft gear
will absorb that much energy during a collision—equal to about 1% of
the kinetic energy of a loaded freight car traveling at 20 mph (32 km/h).

A TRAIN IS STRETCHED IN TENSION when pulled uphill by the locomotives. Going downhill, the train is compressed when braking is concentrated at the front. Things get complicated if a very long train tra-
verses two or more hills with different parts of the train going uphill and
downhill. When a train is going up and down multiple hills, the slack
running in or out creates additional train forces. Different parts of the
train can also compress or stretch when empty cars brake faster than cars
with loads. A long train can be difficult to control if the slack is constantly
running in and out.

Too much compressive force (called buff force) can derail a train by
buckling or jackknifing cars on the outside of a curve. Too much tension
(called draft force) can cause a train to derail by stringlining (explained
later in this chapter) on the inside of a curve (Figure 7.10). Excess tension
can also break a coupling. Buckling on curves is more common.

Slack running in adds additional train compressive forces; slack running out adds additional tension forces. The draft gear absorbs some of
this slack action. Eventually, the draft gear runs out of travel, and the
resulting hard, metal-to-metal impact significantly increases in-train
forces. It's like being hit in the head with a hammer. The impact force
from the hammer is much lower if a spring cushions the blow. The
spring's cushioning effect stops when the spring compresses solid.

Because of the somewhat unpredictable stick-slip nature of the fric-

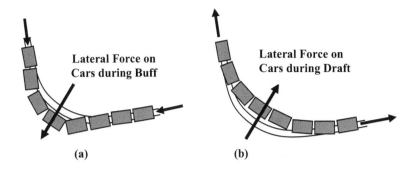

FIGURE 7.10. (a) Too much train compression buckles the train off the outside rail, and (b) too much pulling force in the train derails the train on the inside rail by stringlining.

tion elements, slack action can even be dangerous in a parked train. A worker can be crushed if he or she happens to wander between un-coupled cars when the slack unexpectedly runs out. The danger is well known. Workers are still tempted to walk between uncoupled cars if the alternative is walking 0.5 mile (0.8 km) around the end of the train to safely cross over. Standard rules require uncoupled cars to be at least 50 feet (15.2 m) apart before workers are allowed between them.

EMPTY CARS AND CURVES

An empty car has a lower vertical load V and a higher L/V ratio; thus, it has a greater tendency to derail. When a train is compressed on a curve, the couplings are angled, adding additional lateral force try-ing to shove cars off the outside rail. Short cars attached to long cars create larger coupling angles on curves. The most likely cars to derail on a sharp curve are short, empty cars attached to longer, loaded cars.

Controlling the Slack

The transition from power to braking (or from braking to power) causes in-train forces as the slack moves in and out. The engineer must monitor (and juggle) the electric motor loading, dynamic brakes, engine throttle, and airbrakes for both the main locomotive and any helper engines. Abrupt changes cause problems. The engineer must act slowly to allow the slack to gradually adjust (i.e., allow the draft gear to move to its neutral position). Trains are particularly susceptible to a slack run-in derailment during emergency braking. The most advanced controls automatically trigger the emergency brakes in the back of the train to minimize slack run-in.

Most locomotives used to have only an ammeter to gage the output of the electric motors. Higher amps correspond to more effort from the motors—either more tractive force pulling the train or more dynamic braking slowing the train. Dynamic brakes, discussed in Chapter 10, convert the locomotive's electric motors into generators that slow the train. Concentrated at the front of the train, dynamic brakes run in the slack.

Because the motor current changes with speed, the ammeter is an imperfect indicator of braking forces. Since 2006, all new (and rebuilt) locomotives must provide a readout in pounds of dynamic braking forces.

The train crew mostly maintains control by the seat of their pants (literally) as they scrutinize every little tug based on their many years of experience and training. The train crew also considers hills, curves, and train makeup.

Uphill, Downhill, and on a Curve

An 8,800-foot (2,682-m)-long train weighing 9,440 tons (8,564 MT) with 136 cars (85 empty) derailed in Canada in 2000. Pulling the long train ran out the slack and stretched the train. The front end began to slow as it climbed a hill. Meanwhile, the rear of the train was still accelerating downhill and ran in the slack. The two opposing motions met in the middle, buckled, and derailed 19 cars on a curve at 41 mph (66 km/h).

Railroads prefer to control a train's speed and slack action by raising and lowering the throttle as needed. This optimizes fuel consumption and minimizes wear on the brakes. The next preferred control method is dynamic braking, which also reduces brake wear. Controlling slack with dynamic brakes concentrated in the front of the train, however, needs constant adjustment when the train's speed changes going up and down hills.

The least preferred method of controlling slack is called power-breaking or stretch-braking. With stretch-braking the airbrakes are applied along the length of the train while the locomotive is simultaneously pulling on the front of the train. It's like stepping on the gas while applying the brakes. Needless to say, this wastes fuel and wears out the brake shoes. A locomotive might burn 5 gallons per hour (gph) (18.9 liters per hour [lph]) idling, 14 gph (53 lph) during dynamic braking, and 100 gph (378 lph) during stretch-braking. When cabooses were widely used, stretch-braking was required for the comfort and safety of the crew in back.

The investigators concluded stretch-braking would have been the only way to prevent the violent slack run-in that derailed this long train.

Braking Sequence and Derailments

In 2006, a train more than 8,500 feet (2,590 m) long trailing 8,357 tons (2,547 MT) derailed 20 cars in Virginia. Investigators concluded that improper train-handling on a curve at 29 mph (47 km/h) caused the derailment.

The train was stretched while accelerating downhill. With the train fully stretched and still going downhill, the engineer applied full dynamic brakes at the head of the train. The slack ran in and derailed the train.

The engineer should have gradually applied the dynamic brakes while slowly releasing the airbrakes to prevent the train from fully stretching out. Instead, after cresting the hill and completely releasing the airbrakes, the engineer applied maximum dynamic brakes in just 58 seconds. The slack ran in 13 seconds later. The recorded speed increased 3 mph (4.8 km/h) in just 2 seconds, indicating a severe slack run-in that derailed the train.

Computer simulation estimated a compressive buff force of 288,000 lbs (1,281 kN). This compressive force buckled and derailed a lightly loaded 89-foot (27-m) car on a 1,005-foot (306-m) radius curve.

Train makeup also contributed to the accident. The first 28 cars were lightly loaded, 89-foot (27-m) cars with end-of-car cushioning devices that increased slack action. More than 6,000 tons (5,443 MT) trailed behind this block of lightly loaded, long cars.

According to computer simulation, if the heavier cars had been placed in front, the compressive load reduces to a safe 200,000 lbs (889 kN), acceptable for the conditions. Gradual dynamic braking and airbrake release further reduce the compressive forces to 110,000 lbs (489 kN).

Empty cars should be located in the back of the train to prevent heavier cars in the back from squeezing off the lighter cars up front. All railroad companies have rules about how to make up trains reflecting their operating conditions. Usually, railroads prefer to make up trains based on the next delivery to minimize switching cars. The first cars to be dropped off, empty or full, are placed at the end of the train.

Increased switching activity is also an opportunity for increased worker error, accident, and injury; therefore, it is not immediately obvious that all trains should be made up to minimize in-train forces instead of reducing switching activity. Nevertheless, some companies are beginning to evaluate train makeup with computer programs that analyze in-train forces, especially on particularly difficult routes.

Computer Simulation

Computer programs have been developed to identify safe operating conditions on particularly difficult routes and for forensic analysis of derailments. The computer programs evaluate in-train forces from throttle and braking activity recorded by the event recorder, train resistance, curves and hills, and slack action from the draft gear. All this information is related by calculating $f = m \times a$ for each car. The software calculates the lateral to vertical load (L/V) ratio at the wheel-rail interface on every car along the route, and considers the changing coupler angles for cars on curves, empty cars, and short car–long car combinations.

The rail is periodically inspected with a geometry car that verifies the rails are within spec and records exact rail geometry suitable for input into the computer simulation. The computer software can then calculate changing L/V ratios for warped or shifted rail.

Too Much Dynamic Brakes

Dynamic brakes (DB) concentrate the braking force at the head of a train and increase the tendency to compress and buckle the train. (Recall that each powered axle has an electric motor attached that converts into a generator that slows the train.) There is a practical limit to the maximum DB at the front of the train. Railroads usually describe the limit in terms of number of powered axles allowed. An axle count is used because individual axles can be shut off or individual motors may be defective. The most common configuration is six powered axles per locomotive.

A common limit for general freight trains is 20 axles of DB. Each axle provides 10,000 lbs (44.5 kN) of braking force for a total dynamic braking force of 200,000 lbs (890 kN). For heavy grain or coal trains, the limit might be raised to 24 axles, or 240,000 lbs (1,067 kN). Different limits may be placed for various reasons on specific routes (e.g., use less than 40,000 lbs [178 kN] of dynamic braking on all sharp turns with a speed limit of 25 mph [40 km/h] or less).

But not all axles of dynamic braking are created equal. Different types of locomotives have different types of computerized traction control to minimize wheel slip and increase traction. Slip control also improves dynamic brakes. To account for reduced slippage, axles with better control have their braking force increased by a multiplier when making the axle count. Common multipliers are 1.33 and 1.5.

In 2006, a 7,185-foot (2,190-m)-long train with 3 locomotives (18 powered axles) weighing 11,685 tons (127 cars, 34 empty) derailed going downhill on a 3-degree curve in Tennessee. Even with maximum dynamic braking, the train's speed began to increase to 23 mph (37 km/h). The engineer added airbrakes to slow the train. After the train slowed, he released the airbrakes but left the dynamic brakes on. Thirty-three cars derailed at 22 mph (35 km/h).

The Federal Railroad Association (FRA) ruled that this train derailed because the 24 effective dynamic brake axles exceeded the maximum of 20. The train crew lost track of the correct multiplier needed for a correct axle count.

Distributed Power

As freight trains became longer and heavier, breaking couplings, stringlining, excessive dynamic brakes, and all the other handling problems discussed in this book became practical limits on the number of locomotives placed in the front. Distributed power, that is, using additional locomotives in the middle or end of the train, became the solution.

Railroaders have long used helper locomotives in the back of trains climbing steep grades. Before radio, the two crews coordinated their efforts with whistles. In the 1960s, radio control of helper locomotives began. The big advantage was reduced labor costs. Also, with all the power at the head, stringlining caused increased friction on curves. Distributed power reduced friction and fuel consumption by up to 5% to 10%, improved handling of brakes, and reduced slack.

Integrated distributed power, with alternating current (AC) traction motors, was the next big innovation in the mid-1990s. With integrated distributed power the controls of all locomotives appeared on one computer screen in the head locomotive. Two sets of locomotives coordinated their efforts to reduce in-train forces. Better traction control of the AC motors also reduced the jerky wheel slip starts typical of direct current (DC) motors at that time. One operator reported that the incidence of broken couplings dropped from 7% for their heavy coal trains to 0.5% when using integrated distributed power.

Integrated distributed power can be operated in two modes, either synchronous (all commands are mirrored in the remote locomotive) or independent (each locomotive receives separate commands). After cresting a hill, the locomotive engineer might have the lead locomotive increase dynamic brakes while the rear locomotive pushes to keep the train bunched up.

The greatest ever application of integrated distributed power occurred in Australia in 2002. A 12-mile (19.3-km)-long, 100,000-ton iron ore train with 682 cars was pulled by eight 6,000-horsepower (4,474-kW) locomotives for 447 miles (719 km). The train was configured with 2 locomotives in the front followed by 168 cars, 2 locomotives, 168 cars, 2 locomotives, 168 cars, 1 locomotive, 178 cars, and 1 locomotive at the end. Essentially a publicity stunt, the feat was never attempted again. It deserves all the favorable publicity it can get. Only one coupling broke in route.

Today some operators use distributed power; others use helper locomotives with an operating crew. It depends on their operating conditions and ability to finance new equipment. Canadian Pacific operates 14,000-foot (4.26-km) trains with distributed power, and expects to operate longer trains in the future.

Stringlining

Place a 6-foot (1.8-m)-long chain on a flat table in an arc. Pull on both ends of the chain and it straightens. That's all there is to stringlining. Pull on a train on a curve with too much force, and the train stringlines and derails off of the inside rail (see Figure 7.10). Stringlining forces can also cause the inner rail to roll over.

In 2005, a northbound 9,340-foot (2,846-m) train stringlined and derailed at 14 mph (22 km/h) on a sharp curve in British Columbia. The train, operating with integrated distributed power, had 141 empty cars, 3 loaded cars, and 7 locomotives.

The train was stretched across 9 curves and ascending a 2% grade. Half of the curves were 9 degrees or greater (radius of 636 feet [193 m] or less). The derailment occurred on a 12-degree curve with a radius of just 478 feet (145 m). This very difficult hilly, curvy route had its own special operating rules.

This train had 5 locomotives in the front and 2 remotely controlled in the rear. Of the 5 locomotives at the head of the train, one was shut off and a second was not electrically connected; there were only 3 locomotives pulling with 11,800 horsepower (8,800 kW) and 18 driving axles.

The remote locomotives in the back were accidently set to pull in the wrong direction. Protective circuits automatically shut off the incorrectly set motors. Because the control logic for the front and back locomotives were of different vintages, the rear locomotive's shutoff alarm was not communicated to the train crew's locomotive. If all the alarms had been properly working, the train would have never left the yard.

From the train's response the crew quickly figured out that the locomotives in the rear were not working. As soon as the remote locomotives failed, the train became a conventional train subject to very specific operating rules unique to this route with its sharp curves and steep hills.

Operating rules for trains without distributed power on this difficult route have been repeatedly changed, most recently in 2003. The review included computer simulation and test train runs. Trains without distributed power were limited to 13,500 horsepower (10,067 kW) at the head of the train. Also, the trailing weight behind empty cars longer than 80 feet (24 m) was limited to 2,700 tons (2,449 MT) and the number of empty cars was limited to 80. Just a few months later, the operating rules for this route were revised again.

Instead of a limit on horsepower at the head of the train, new limits were placed on the number of powered axles: 12 axles for trains with all empty cars and 18 axles for trains with empties and loaded cars not exceeding 4,600 tons (4,173 MT). The limit was changed from horsepower to the number of axles because excess pulling force, not horsepower, stringlines trains.

Because of reduced slack action, trains with distributed power had no limits. As soon as the train in this accident lost power from the remotely controlled rear locomotives, it changed from being a distributed power train with no restrictions to a conventional train with severe restrictions.

Without the two locomotives in the rear, the three active locomotives in the front did not have enough traction to pull the long train up a 2% grade. When the train's speed slipped from 25 mph (40 km/h) to 8 mph (13 km/h), the crew turned on the fourth locomotive at the front of the train. There were now 16,200 horsepower (12,080 kW) and 24 driving axles at the head end.

Stalling a train on a steep hill is an operational if not a potential safety problem. Given the circumstances, the investigators considered turning on the fourth locomotive to be a reasonable action. Besides, at this point the crew's responsibility is to safely operate the train, not sort out the correct combinations of train operating rules.

Adding the fourth locomotive created too much pull and stringlined empty cars on the sharp curve. The combination of excess pulling force in the front of the train, long car–short car combinations, and empty cars created L/V ratios that were too high on the curve.

After the accident, the operating rules on this route were changed yet again. Distributed power trains were limited to 99 cars and 6,000 tons (5,443 MT) maximum. No more than 3,750 tons (3,402 MT) can trail an empty car 76 feet (23 m) or longer. All trains are limited to a maximum of 3 locomotives, with 2 in the front and 1 remote in the rear. Operation of the remotely controlled locomotives must be verified before leaving.

Two other much shorter trains (80 cars and 90 cars respectively) derailed by stringlining on these tight curves. In both cases the emergency brakes near the end of the train were triggered for unknown reasons. Pulling in the front and braking in the rear stringlined both trains.

Stringlining can also occur if the engineer applies the throttle too quickly after releasing the brakes. When releasing the brakes it takes time for the airbrake signal to reach the back of the train. If the engineer releases the brakes and then applies the throttle before the entire train has released its brakes, the front of the train is pulling while the back is still braking—ideal conditions for stringlining on a curve. (It takes 2 minutes to completely release the brakes in a 100-car train, 4 minutes for a 150-car train.)

Another practical limit on the locomotive pulling forces at the head end is the strength of the couplings. The couplers must provide the force to pull the train up hills and the force to accelerate the train per $f = m \times a$.

The couplings must also withstand shock loads. If the wheels slip and then suddenly grab, the whole train is jerked with a shock load. Just as a hammer blow to the head is different from gently placing the hammer on your head, a dynamic shock load is considerably different from a static load.

Couplers come in two breaking strengths: 350,000 lbs (1,557 kN) for general use and 650,000 lbs (2,891 kN) for bulk commodity trains. To provide a margin of safety, railroads de-rate the breaking strength for everyday use based on their experience with breaking couplings. Example values used are 280,000 lbs (1,245 kN) and 390,000 lbs (1,735 kN).

Rail Rollover and Train-Handling

In 2001, a Canadian train with 4 locomotives was hauling 86 loaded grain cars weighing more than 11,000 tons (9,979 MT). The 5,355-foot (1,632-m)-long train derailed 59 cars on a 4-degree (1,432-foot [436-m]) radius) curve at 64 mph (103 km/h). The speed limit on this track is 60 mph (97 km/h), reduced to 50 mph (80 km/h) on the derailment curve.

Earlier in the trip the train had experienced a "kicker" (an unexpected emergency brake application) after applying the airbrakes. After the crew stopped the train for inspection and found nothing obvious, the train proceeded—not an uncommon occurrence. Fearing further delay and potential derailment from unwanted emergency braking, the crew decided to avoid any further airbrake use and complete their journey with only dynamic brakes.

Of the 4 locomotives, one had no dynamic brakes. Confused about the correct multipliers for the different type of traction control, the train crew thought the 3 locomotives had 20 effective axles of dynamic brakes. Because the operating rules for this route called for a maximum of 18 effective axles, the engineer shut off the dynamic brakes on one locomotive. What the train crew thought were 14 effective axles of dynamic brakes turned out to be only 12.5. Without using airbrakes, 12.5 axles of dynamic brakes were not enough to handle this train on that hill.

About 2 miles (3.2 km) from the start of a 7-mile (11.2-km) descending hill, the train was on level track and moving at 49 mph (79 km/h) in full dynamic brakes. Downhill the train accelerated to 64 mph (103 km/h) and derailed after overturning a rail.

Because of the "kicker," the engineer did not want to apply the airbrakes. He rightly feared emergency brakes could derail the train. Instead, he decided to ride it out and try applying the locomotive independent brake.

If the wooden cross ties are weak, the distance between rails will increase on curves. The centrifugal inertial loading will incrementally shift the outer rail away from the inner rail.

The distance between rails for this Class 4 track (called track gage) is between 56 and 57.5 inches (1.42 and 1.46 m). Official track gage in North America (and much of the world) is 56.5 inches. The ranges presented here include the tolerances for the different classes of track.

Class 4 track is limited to 60 mph (97 km/h) maximum for freight trains. When track gage is wider than 57.5 inches, a "slow order" is issued. Until repaired, the track is derated to Class 3. Class 3 track, restricted to 40 mph (64 km/h), can be 0.25 inch (0.63 cm) wider.

About 6 months before the accident, track inspectors found 16 locations within 0.25 inch of the cutoff requiring a slow order. Three months before the accident, 6 sites measured just 0.125 inch below the limit. At that time the track was adjusted to 57.25 inches to buy a little more time. This section of track was working its way to the top of the "urgent repair" list.

A major railroad operating tens of thousands of track grades its track (to prioritize repairs) based on a variety of criteria. Track used by passenger trains or freight trains hauling hazardous material will receive a higher priority.

In another modern development, most main line track is monitored with a gage restraint measurement system, or GRMS. While traveling along the track, the GRMS system hydraulically spreads the rail with a force of 14,000 lbs (62 kN) to assess the wooden cross ties and anchoring system. The track inspectors extrapolate the gage widening at 14,000 lbs to 24,000 lbs (107 kN), the expected force if a fully loaded train makes an emergency stop. This track passed the GRMS lateral load test six months before the accident.

After the accident, wide-gage sites noticed 6 months earlier (and not destroyed by the accident) were reexamined. Some of them had grown to 0.19 inch (0.5 cm) past the cutoff requiring the speed reduction. Also, 35% of the ties had rotted to the point of not being able to hold their spikes, including a cluster of 5 at the derailment site. Surprisingly, the rules only require 12 non-rotten wooden cross ties for every 39 feet (11.9 m) of Class 4 track and only 8 good cross ties for Class 3 track. (An example cross tie spacing is 24 every 39 feet.) The rules also require maintaining a safe track width.

Computer analysis after the accident predicted the track lateral load on this curve for this train at 50 mph (80 km/h) was 13,350 lbs (59 kN). At the derailment speed of 64 mph (103 km/h), the lateral load increases to 20,500 lbs (91 kN), apparently a load too high for this slightly wide track with questionable wooden cross ties. The outer rail rolled over. For a train without airbrakes, the computer predicted the hill needed to be crested at 30 mph (48 km/h) or less to stay below the 50 mph (80 km/h) speed limit on the curve.

A combination of circumstances derailed this train. If the train had left with more dynamic brakes, it could have safely negotiated the hill. If the train had not had an unexpected emergency brake application, the train crew would have used the airbrakes. If the cross ties had been in better shape, the rails would not have spread to a marginal limit. It also would have been helpful if track maintenance had occurred sooner. After the accident, the operating rules increased the maximum number of dynamic brake axles from 18 to 20 on this route.

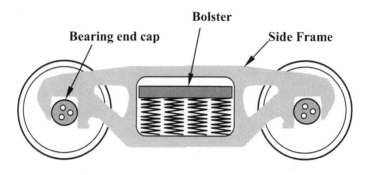

FIGURE 7.11. Side view of a single freight car truck assembly.

Bounce

Freight cars sit on two trucks, one on each end. The truck supports the wheels, axles, and bearings. The standard three-piece truck consists of two side frames and a spring-mounted bolster. The bolster sits on springs mounted in the side frames (Figure 7.11). The side frames grip the bearings on each axle. (Also, see Figures 8.14 and 8.15 in Chapter 8.)

Just as a car on a bumpy road will bounce the passengers up and down, a train traveling on bumpy track will bounce the freight car body up and down on the spring-mounted bolster. A spring-mass system bounces up and down with a special rhythm known as its natural frequency.

One way to measure the natural frequency of weight sitting on a spring is to press down on the weight, release, and count how often the weight bounces up and down during a period of time. Example values for a 286,000-lb (129,700-kg) freight car might be about 2 cycles per second (cps) when loaded and 4 to 6 cps when empty. If the natural frequency of the freight car happens to coincide with the frequency of dips in the rail, then a special condition known as resonance may occur.

Resonance is easier to describe with a child's swing—a pendulum. If the swing is pulled back and released, it swings back and forth at the pendulum's natural frequency. Let's say the pendulum swings back and forth two times per second and therefore has a natural frequency of two cycles per second. If one stands behind the swing and gives it a tiny push timed at the swing's natural frequency, and if the push always coincides with when the swing is just about to reverse itself and swing down, each push will add to the swing's motion. The swing's motion increases every cycle and is resonating.

During resonance the motion of the swing keeps increasing until something gives, usually the pusher's or pushee's nerve. When mechanical systems resonate, it's a bad thing. The motion continues to increase until something breaks.

The natural frequency of a freight car is not fixed. It will vary with the weight of the freight and the condition of the car's springs. The springs can wear out from too much bouncing, or they can be stretched or sprung from rocking. If the downward, upward, downward, upward motion of the rail dips happens to coincide with the frequency to which the freight car wants to naturally bounce up and down, the two will reinforce each other and resonance occurs, just as with the child's swing.

Freight cars that bounce are not a good thing. If the car bounces too hard, it can literally bounce off the track. The freight car bouncing up and down reduces the weight on the wheels and increases the L/V ratio and the likelihood of derailing.

Just such a derailment occurred in 2006. A coal train with 2 locomotives in the front, 122 cars loaded with more than 17,000 tons (15,422 MT), and a third remotely controlled locomotive in the rear derailed 40 cars on straight track at 50 mph (80 km/h).

Coincidentally, this track had been inspected by a geometry car just 3 weeks earlier. The data showed 3 consecutive track dips of only 1 inch (2.54 cm). Computer simulation of this combination of track dips, train speed, spring stiffness, and weight showed the dips were spaced and tuned such that they unloaded the wheels from about 37,000 lbs (164 kN) per wheel to almost nothing. A different combination of speed, dips, and weight, and the train would have safely passed.

Track dips alone are not expected to bounce a car off the tracks. The bolsters have spring-loaded friction wedges (similar to draft gear) that resist excess bounce. The friction wedges act like shock absorbers on a car. But, of course, the friction wedges can wear out. Modern track sensors that detect high-impact wheel loads can also identify cars with worn-out suspensions that excessively bounce.

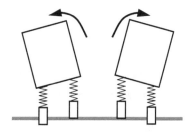

FIGURE 7.12. A freight car can resonate by rocking back and forth.

Harmonic Rock-Off

Harmonic rock-off can derail a train that is moving too slowly. Before continuously welded rail was widely used, the rails had bolted joints every 39 feet (11.9 m) (see "Rail Joints" in Chapter 12). The rail joint is a natural weak point that flexes and causes a wheel to dip. Because the joints are staggered, the wheels dip one at a time, first on the right side, then on the left side, causing the freight car to rock back and forth, as shown in Figure 7.12. (Freight cars can also resonate with a pitching or yawing motion.) If the motion magnifies or resonates, the wheels can lift off. The friction wedges that resist bounce also resist rocking. Worn-out friction wedges are usually the primary cause of harmonic rock-off. Freight cars with a high center of gravity and worn-out friction wedges are most susceptible.

I remember watching a slowly approaching freight train with rocking box cars as a child in the 1960s. We dared ourselves to stand perhaps 10 feet (3 m) from the tracks. It was absolutely terrifying to watch the cars rock like a massive wall about to tumble on top of us. We all lost the dare and ran away.

TRUCK-HUNTING

A freight car sits on two trucks that rotate under the car body on curves at the center plate (Figure 7.13). The center plate on the bolster is connected to a center plate on the freight car's bolster with a center pin.

Because the wheels are fixed on the axle, both wheels must rotate at the same revolutions per minute (rpm). On a curve, the outer wheel travels a farther distance than the inner wheel. If both wheels are fixed on

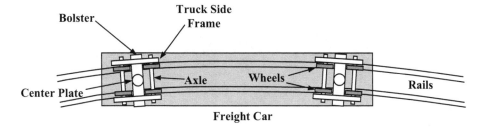

FIGURE 7.13. A freight car with two trucks rotating underneath the car body on a curve (lying on rails looking up).

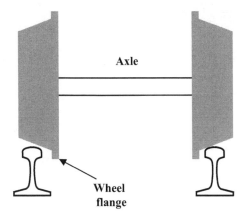

FIGURE 7.14. Railroad wheels are conical. The cone angle shown (front view) is greatly exaggerated.

one axle and forced to travel two different distances, one of the wheels must slide. A car solves the same problem with a differential that allows both wheels to turn at different speeds on a curve.

On rail cars, the curving problem is addressed by using conical wheels (Figure 7.14). The outer wheel on a curve will automatically move to a larger diameter and the inner wheel to a smaller diameter. This allows both wheels to travel different distances while rotating at the same speed and minimizes wheel sliding, rubbing, and wear. Theoretically, the wheels should center themselves on a curve and not rub on the wheel flanges. Unfortunately, wear of the rails and wheels quickly alters the ideal state. Wheel flanges do rub on curves, and curved track wears out faster than straight track.

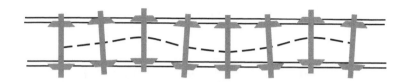

FIGURE 7.15. Truck-hunting. The truck oscillates from side to side every 30 to 50 feet (9 to 15 m).

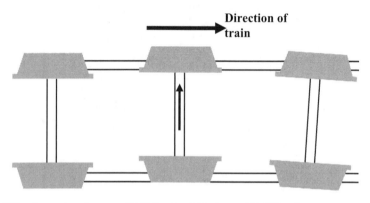

(A) Wheels centered   **(B) Wheels shifted**   **(C) Wheels steer to re-center**

FIGURE 7.16. After rolling over a misaligned rail the conical wheels are designed to automatically center the wheels on straight track.

On straight track the conical wheels should remain centered on the rails. The track is not always perfectly straight, however. Crooked rails shove the wheels sideways. The wheels then "hunt" back and forth on unequal diameters, trying to find a balance (Figure 7.15).

Figure 7.16 illustrates how the wheels re-center themselves on straight track. After rolling on a patch of crooked track, the axle shifts as shown. Even though the two wheels are rotating at the same rpm, the top wheel rim's speed (now rotating on a larger diameter) is faster than that of the bottom wheel. The axle steers itself to the center of the track as shown.

Truck-hunting occurs on straight track above some critical speed. Below the critical speed, friction dissipates truck-hunting. Above the critical speed, the self-steering is too far in one direction and requires a rapid correction in the opposite direction. As the speed increases, the self-steering correction increases and truck-hunting amplifies.

Eventually, the wheel flanges can alternately strike the two rails. The impacts can damage the rails, wheels, trucks, and cargo. If the wheel flange impacts continue to grow, a wheel climb derailment can result. Fifty-two derailments from truck-hunting were recorded in a recent 10-year period.

Truck-hunting can interact with other factors. In 2004, a Canadian train derailed 27 cars when traveling at 58 mph (93 km/h). The derailment was caused by wheel lift during truck-hunting after a dynamic brake–induced slack run-in.

TRUCK-HUNTING: EXPLAINED AGAIN

Consider a ball thrown at a very shallow angle against a wall. The ball will bounce off in a glancing blow. If there is another parallel wall close by, the ball will bounce into the second wall (Figure 7.17). The harder the ball is thrown, the more times it will bounce off the two walls. Eventually, the deformation of the ball and wall during each impact dissipates the ball's kinetic energy of motion—the ball stops bouncing.

If the ball is propelled by a small rocket engine (or pulled by a train) and is supplied with additional energy, the ball will continue bouncing between the two walls. If the added energy is greater than the energy dissipated by each bounce, the ball will bounce between the two walls indefinitely—the critical speed has been reached.

If the ball's speed increases, the bouncing occurs with greater impact force. Self-steering wheels and/or the wheels' flanges impacting on the rail provide a rotation that changes the direction of the axle analogous to the ball bouncing off the walls.

Empty freight cars may be susceptible to truck-hunting at speeds of 45 to 55 mph (72 to 89 km/h). In one test a flat car in a "worn" condition and with a load capacity of 70 tons (64 MT) began truck-hunting at 45 mph. Renewing the worn truck suspension increased the truck-hunting

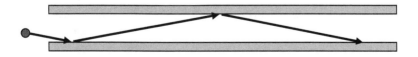

FIGURE 7.17. A ball thrown at a shallow angle bounces between two walls.

speed to 65 mph (105 km/h). Using constant contact side bearings (CCSB) also increased the critical speed of this car to 65 mph.

CCSB attach to the truck's bolster and press on the bottom of the freight car body, creating a frictional resistance to truck rotation. CCSB have been required on new freight cars since 2002. Unfortunately, increased friction from CCSB makes truck rotation on curves more difficult, leading to increased fuel use and rail wear.

Truck-hunting is also very sensitive to wheel profile. The perfect wheel profile (that optimizes hunting, curving, and wear) is another area of ongoing research. In one study improved wheel profiles raised the critical truck-hunting speed on an empty car from 56 mph (90 km/h) to 80 mph (129 km/h).

Just as a higher-speed ball will bounce more between the two walls, truck-hunting is particularly severe for high-speed passenger trains. Rail straightness standards on high-speed tracks are set, in part, to limit truck-hunting. In fact, all trucks on all trains operating above 125 mph (201 km/h) have sensors that sound an alarm if truck-hunting occurs.

To eliminate truck-hunting Amtrak's 150-mph (241-km/h) Acela attaches two hydraulic dampers, similar to a car's shock absorbers, to the heavier locomotives on each end of the train. The 200-lb (91-kg) damper

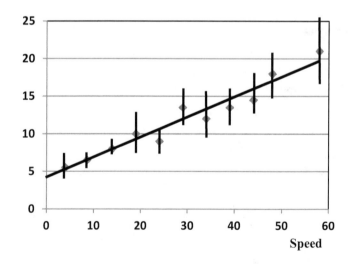

FIGURE 7.18. Relationship between derailment speed and number of cars derailed. *Adapted from Barkan et al.*

is connected to the locomotive's frame with steel brackets about 16×20 inches (40.6×50.8 cm) and 0.75 inch (1.9 cm) thick. The Acela dampers made headlines in August 2002, when fatigue cracks were found in the connecting brackets. All Acelas were grounded until thicker brackets could be installed.

Because there are now 70 truck-hunting detectors (THDs) installed on higher-speed freight train tracks, truck-hunting accidents are trending down significantly. The THD measures forces on a rail similar to a high-impact wheel detector (see Chapter 11).

DERAILMENT SEVERITY

More cars are likely to derail with increased speed, as shown in Figure 7.18. The figure is from a study of hazmat accidents. A heavy coal train would derail less; a load of empties would derail more.

8

Bearing Failures

The Congressional Limited was the crown jewel of America's foremost railroad, the Pennsylvania Railroad company.[1] Traveling from Washington, D.C., to New York City, mostly at 70–80 mph (113–129 km/h), the Congressional Limited covered the 220 miles (354 km) in just 3.5 hours.

Shortly after six o'clock p.m. on September 6, 1943, railroad yard workers spotted a bearing fire on the 7th car of the 16-car train. They tried contacting the next signal tower, but it was too late. After slowing to 56 mph (90 km/h) on a curve, the train derailed about 1 minute later.

When a bearing fails by overheating, it can quickly burn off the axle. Apparently, the dragging undercarriage speared a wooden tie and catapulted the seventh car nearly straight up, dragging the eight cars behind it off the tracks. The seventh car, with the eighth car still attached, repeatedly rolled before striking the concrete base of a signal tower. The impact tore the seventh car almost its entire length. The eighth car wrapped around the tower roughly in the shape of a U. Almost all fatalities were in Cars 7 and 8.

When the accident was announced on local radio, 40 priests quickly arrived. The accident scene was covered with dazed men, women, and children. Many of the estimated 250 passengers on the derailed cars had

FIGURE 8.1. Wreckage of the Congressional Limited. *Department of Public Safety, Philadelphia.*

been hurled through windows. Ambulances were still bumper-to-bumper six hours after the wreck. For extraction from the wreckage, at least two victims had their legs amputated (Figure 8.1). Eighty passengers would die immediately or within the next few days.

Procedures call for bearing inspections at every opportunity. On curves trainmen are taught to lean out and look back along the line of cars. The crew testified that a visual inspection had been done 5 miles (8 km) before the derailment. A mechanic had inspected the bearings up close just hours before the accident.

By 1943, many passenger trains had switched to safer roller bearings. In fact, the Congressional Limited had roller bearings on all cars, except two older cars (including the doomed seventh car), added for increased wartime traffic (Figure 8.2). Angry editorials denounced the older

FIGURE 8.2. Intact wheel and axle (*above*); the missing bearing is mounted to the left of the wheel. Burned-off axle that caused the accident (*opposite*).

bearings. The Pennsylvania Railroad tried to explain the burdens of the war effort.

Military troop movements and civilian gas-rationing combined to create the largest passenger traffic in American history.[2] Between 1939 and 1942, the number of passengers and the amount of freight increased by 234% and 90%—the workforce by only 39%. Freight and passenger traffic grew again by 48% and 15% in 1943, without any added workers. Meanwhile, the turnover in manpower was tremendous (don't forget the army, navy, and marines). There were 820,000 new railroad employees in 1942 alone. The railroads were operating aging, even retired rolling stock on crowded schedules with inexperienced workers. Also in the news in 1943, described as the biggest air battle of the war, was a bomber raid on a German ball bearing plant.

RAILROAD BEARINGS

A primitive journal-bearing in ancient times was simply a greased axle turning inside a hollow cylindrical hub or sleeve. Various combinations of wood-metal axles and sleeves were used from ancient times. Before the widespread use of roller bearings, a railroad plain journal-bearing was more or less a metal-on-metal version of a greased wagon axle.

A railroad car sits on two trucks. Each truck has two axles with two bearings on each end for a total of four bearings per truck, eight bearings per car (Figure 8.3). The wheels are pressed on the axle (i.e., rigidly attached). The axle/wheels rotate as a single piece inside the bearing box.

Bearing Box

FIGURE 8.3. A truck displayed at the Illinois Railway Museum (circa 1920s).

For the older-style plain journal-bearing, which was used on the Congressional Limited Car 7, the weight of the car sits on brass wedges. The wheels rigidly attached to the axle rotate underneath the wedges. The end of the axle that rotates underneath the wedge is called the journal. The journal is machined with a very smooth finish. This configuration has been used relatively unchanged since 1830 (Figure 8.4). Standard journal sizes adopted in 1920 are shown in Table 8.1. A 6.5 inch × 9 inch (16.5 cm × 22.8 cm) journal is used on today's 286,000-lb (129,730-kg) freight cars, albeit with stronger, modern steel.

To remain lubricated the plain journal rotated through oil-soaked, loose cotton rags (known as packing) in the bottom of the bearing box. A tangled mass of loose threads (called waste) salvaged from textile mills was commonly used. Long, intertwined strands were best. Loose threads could get stuck between the brass and journal (known as waste grabs), disrupt lubrication, overheat, and ignite. Bunched-up packing would glaze, limit lubrication, and create hot spots. Packing rolled up on one side could lead to waste grabs. The packing also had to be frequently "dressed." Dressing involved pulling out waste grabs, unbunching, and rotating the packing. Because the bearing box was not well sealed, oil leaked out and dirt got in. The box had to be checked daily for dirty or low oil, heat damage, and glazed or bunched packing.

FIGURE 8.4. Bearing box, brass wedge, and journal.

TABLE 8.1
Standard Journal Sizes Adopted in 1920

Journal Size*	Axle Capacity (lbs)	Car Weight (tons)
3.75 × 7	12,500	33
5 × 9	27,000	68
5.5 × 10	34,000	84.5
6 × 11	42,500	105

*The first number is the axle diameter in inches. The second number is the length of the journal.

The packing also filtered the oil and needed to be replaced or "repacked" every 12 months or so. The Congressional Limited bearing had oil added on the day of the accident and had been repacked less than three months before.

Before roller bearings were adopted, the "hot box" was all too familiar to passengers, as a nineteenth-century passenger explains:

The advent of a "Hot Box" was heralded by an "aroma" modest at first but growing more insistent as the train sped along and causing travelers to sniff until an outspoken one called across the aisle to a friend—"Hot Box"? and received a nod in reply. Then came rattling of bell cord and quick answer from the whistle. The train drew to a halt . . . our brakeman (appeared) . . . with a pail of water in one hand . . . and a pail slopping with a dreadful looking mess. His poker scalped the journal lid and exposed the heated parts belching smoke. He hooked out the smoldering waste and dashed water inside which forthwith rushed away in sizzling bursts of steam. Then he poulticed the feverish interior with that dreadful looking dressing poked in by rod and hand.[3]

A 1952 issue of *Railway Age* pointed out that "there is no greater single cause of train delays." It has been often said that the poorest locomotive reliability record in the history of any railroad can hardly equal the train delays caused by the best hot-box record on that same railroad.

Bearing reliability actually got worse in the early 1950s as the diesel engine replaced steam locomotion. The steam engine's frequent stops for fuel, water, and maintenance provided more opportunities for bearing inspection. In 1949–51, a train had to be stopped every 100,000 to 360,000 car miles (160,934 to 579,364 km) (hotter months were worst) to set out a car with an overheated bearing.

In 1946, bearing failures caused about 4% of all derailments (counting human error, track problems, and equipment failures). After dieselization a decade later, the percentage rose to 14%. In 1956, bearings were responsible for 30% of all equipment-related derailments. One must also consider the inconvenience of tens of thousands of delays per year caused by hot bearings.

The spring-loaded pad was developed in the 1950s (Figure 8.5). The pad stayed together better and lasted longer. But excess pad pressure and friction could degrade the lubricant and pad.

In 1980, one railroad company averaged 700,000 car miles (1,126,541 km) between hot boxes for journal-bearings with lubricating pads. While this number was significantly better, roller bearings (discussed later in this chapter) for that same railroad were averaging 16 million car miles 25,749,504 km) before overheating.

Without adequate lubrication metal-to-metal contact quickly gener-

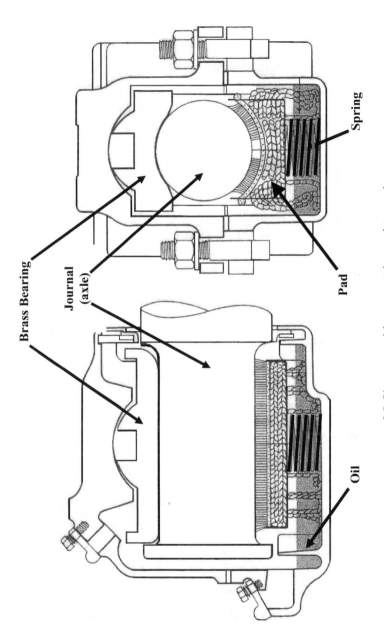

Brass Bearing

Journal
(axle)

Spring

Pad

Oil

FIGURE 8.5. Plain journal-bearing with a packing pad.

ated heat and smoke and ignited the oil-soaked packing. Hence the term *hotbox*, used today to describe any overheated bearing, even though modern roller bearings do not have any actual box. If the train is not stopped soon, the frictional heat will build up and burn off the axle. The steel axle does not actually burn, but overheats, loses strength, and breaks off.

There were other problems with plain journal-bearings. A fire in the hot box could ignite and burn freight. One 3-year period (1948–50) averaged 300 freight cars fires per year ignited by bearings. Needing frequent replacement, loose brass wedges were stored on every train and in every station. The brass had considerable scrap value and kept disappearing. In 1971, the United Kingdom reported 80,000 stolen brass wedges. Even the waste packing would disappear. One railroad division reported more than 100 instances of missing packing—apparently oil-soaked rags were good for starting hobo fires.

HYDRODYNAMIC ACTION

The principle of hydrodynamic action was discovered in the early 1880s by Beauchamp Tower while studying a better way to lubricate railroad bearings. A railroad-style bearing was loaded, as shown in Figure 8.6. When Tower drilled a hole in the bearing's top surface to add more oil, oil squirted out. Mounting a pressure gage in the hole, Tower discovered a pressurized film of oil separating the axle and the brass wedge. He measured the pressure in the film at about 300 to 350 psi (2,068 to 2,413 kPa), or twice the average bearing pressure. (The average bearing pressure equals the load on the bearing divided by the bearing diameter multiplied by the length of the journal.)

It is easier to explain the hydrodynamic action inside a complete 360-degree plain journal-bearing (Figure 8.7). Initially, the axle is at rest. As the axle begins to spin, it tries to climb the wall. As the speed increases, oil adheres to the surface of the axle and is wedged into the gap, creating pressure under the axle. Eventually, the shaft "floats" on a film of oil.

The concept of liquid adhering to a moving surface is counterintuitive, but can be illustrated by slowly pulling a knife from a container of heavy motor oil. Fluids with less internal friction (called viscosity) form a thinner layer on the knife. Even air will adhere to a moving surface.[4]

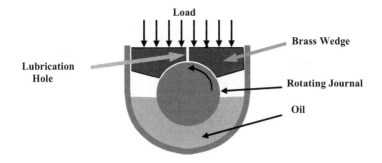

FIGURE 8.6. Experimental setup from 1880 that measured the pressurized film of lubricating oil.

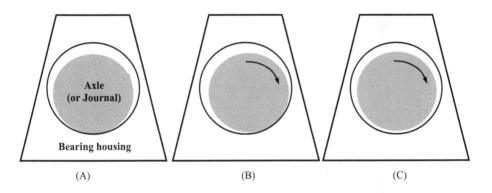

FIGURE 8.7. (A) Axle at rest. (B) Axle starts to rotate clockwise and tries to climb the wall of the bearing. (C) Axle is up to speed. With full hydrodynamic action, the axle "floats" on a pressurized film of oil.

A similar phenomenon, called hydroplaning, occurs under a car tire on wet pavement. As the tire speed increases, the water wedged under the tire collects faster than the tread can funnel it away. Water pressure builds under the tire, and eventually the car floats on a pressurized film of water.

The plain journal-bearing starts out with metal-to-metal contact and very high frictional forces. Just like the hydroplaning car, the railroad axle needs sufficient rotational speed to properly form the lubricating film of oil.

The force to start a train with plain journal-bearings is about 28 lbs per ton of train. (A 100-ton freight car must be pulled with $28 \times 100 =$

2,800 lbs [12.45 kN] of force before it starts moving.) The running friction is about one-tenth this value. Long trains had to be bunched or compressed so that pulling in the opposite direction started up each car one at a time. An additional locomotive was sometimes required to overcome the starting friction. Once the lubricating film was established, the second locomotive could uncouple.

Most of the wear on bearings occurs during startup. A similar problem occurs with the journal-bearings inside a car engine. (Roller bearings cannot be installed on a crankshaft with its rotational offsets.) Stop-and-go driving is well known to cause more wear on an engine than highway driving. For this reason police cars and taxis rarely shut off their engines.

The lubricating film is very thin, typically 0.0003 to 0.0008 inch (0.0076 to 0.020 mm) thick. The axle and bearing surfaces have to be polished as smooth as possible or else the microscopic surface peaks poke through the lubricating film and disrupt the flow.

A break-in period with gentle operations is needed to polish the surfaces to a smoother finish. Without a break-in period, the two surfaces could damage each other. Because train-operating conditions are not easily changed, plain bearings were particularly prone to failure during their run-in period.[5] Low speeds, when the oil film is not properly formed, will cause excess wear. Bearings can also be overloaded. Excess load can push the surface peaks through the thin lubricating film and create metal-to-metal contact.

The journal-bearing will overheat if wear particles (or dirt) accumulate in the lubricant, if there is not enough lubricant, or if the axle is overloaded. Once metal-to-metal contact occurs, the bearing rapidly overheats.

AXLE HEAT GENERATION

Shaking a container of BB's, or ball bearings, for tens of seconds will result in a measurable temperature change. Conservation of energy requires the mechanical energy of motion to be converted into frictional heat energy.

Rapidly bending a paperclip also demonstrates the conversion of me-

FIGURE 8.8. Rapidly bend the paper-clip back and forth 15 times and place it against a dry lip to feel the heat that has been generated.

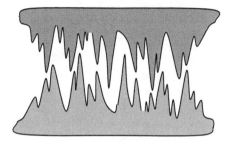

FIGURE 8.9. Microscopic peaks on rubbing surfaces cause bending and tearing that create heat.

chanical energy into heat energy. Use the large end of a "Jumbo" paper-clip (1.8 inches [4.6 cm] long) and bend it rapidly back and forth about ±45 degrees, as shown in Figure 8.8.

The generated heat is very obvious if the paperclip, rapidly bent 15 times or so, is quickly pressed against dry, sensitive skin (the lips are ideal). On an atomic scale, atoms are sliding back and forth, creating atomic-scale friction.

If the paperclip is slowly bent, the heat lost to the air keeps up with the input of energy and little heat accumulates in the paperclip. If the paper-clip is bent rapidly, the heat generated exceeds the heat dissipated and the paperclip begins to build up heat, similar to a railroad bearing and axle during failure. During bearing failure, the extreme heat weakens the steel axle and lowers its life from hundreds of thousands of miles to per-haps just a few.

Bearing failure occurs with violent metal-to-metal rubbing that causes the microscopic surface peaks to weld together, bend, and tear. The welded surface peaks tear out and create new surfaces that perpetuate the process (Figure 8.9). The process of heat generation from frictional rubbing of the surface peaks is similar to the heat generated from bend-ing the paperclip.

BEARING-INDUCED COLLISIONS

In addition to causing spectacular derailments, hot boxes have also been responsible for a few stunning collisions. The first train stops because of a hot box and gets hit from behind by a second train.

On August 9, 1945, a passenger train left Fargo, North Dakota, and stopped repeatedly because of a hot bearing. The bearing had been re-packed in St. Paul, Minnesota, 394 miles (634 km) east of the accident site the day before. The bearings were also inspected by mechanics 54 miles (87 km) before the accident. The mechanics added oil and dressed the packing. The conductor smelled burning oil 12 miles (19 km) before the accident and stopped the train at 6:40 p.m. The hot box was cooled and repacked, and a water-line was rigged to squirt water on the bearing. The train restarted at 7:02 but stopped again at 7:07 to adjust the hose. It restarted again and stopped at 7:18 when the bearing started smoking. The train was now 31 minutes late. The collision occurred 4 minutes later.

The second train was traveling 57 mph (92 km/h) when it saw the stop signal just 1,100 feet (335 m) ahead. Impact was estimated at 45 mph (72 km/h). The rear car was telescoped almost its entire length by the striking locomotive. All 34 fatalities occurred in the telescoped car.

Other collisions with tragic results occurred in 1918 and 1944. On June 22, 1918, a circus train in Indiana was struck from behind just 2 minutes after stopping because of a hot box. The second train, an empty troop train, hit the stopped train at an estimated 25 mph (40 km/h). Four wooden sleeping cars were entirely destroyed. Sixty-seven people died in the collision and subsequent fire. On December 31, 1944, 50 people were killed when a passenger train was struck from behind after slowing to 8 mph (12.8 km/h) behind a freight train with a hot box.

HOT BOX DETECTORS

Until the 1980s, trains were required by law to have a caboose, the last car in the train occupied by the conductor. The caboose served many purposes, including frequent visual inspections of the bearings by the conductor. A plain bearing with its oil-soaked packing could ignite. Overheated grease in a roller bearing would visibly smoke.

Cabooses have been eliminated for a variety of economic and technical reasons. Major reasons are improved reliability of roller bearings and hot box detector use.

Heat Energy

Before the nineteenth century, scientists assumed only visible light (wavelengths between 380 and 760×10^{-9} m) existed. In 1800, a British astronomer aimed sunlight passed through a prism at a thermometer and discovered wavelengths longer than 760×10^{-9} m, called invisible infrared light, or "heat rays."

The heat energy radiated by an object increases as its temperature increases. At around 260°F (126°C) heat can be felt by human skin from an inch (2.54 cm) or so. If the object continues to heat up (and doesn't char), the infrared radiation increases wavelength and transitions to a dull red visible light at around 1,250°F (667°C). With increased temperature, the dull red light becomes the other colors of the visible spectrum. Most of the energy emitted is infrared over a range of wavelengths. An incandescent light bulb, for example, emits 90% of its energy as infrared heat and only 10% as visible light.

An infrared sensor consists of an optical system and a detector. The optics focuses the energy emitted by heated objects onto a detector, a special crystalline transducer[6] that converts heat energy into electrical voltage. The special crystal, called pyroelectric material, converts a changing temperature into a voltage. Hot bearings can be measured from a moving train without physical contact.

Modern railroads install hot box detectors (i.e., infrared sensors) at 20- to 30-mile (32- to 48-km) intervals. If a high temperature is sensed, the sensor sends a radio message to the train (or a dispatcher) identifying the axle number of the hot bearing. The procedure calls for the crew to stop the train, search for the hot bearing, and set out that car on a nearby siding. The hot bearing is verified with a special crayon made of wax that melts at 200°F (93°C).

The hot box detector is not error free. One Canadian railroad's data showed that the crew could not find a hot bearing after an alarm 30% to 40% of the time. By the time the train is stopped and the crew walks maybe a mile (1.6 kg) to locate a hot bearing, it may have cooled off. Rain and driving snow also affect the sensor's ability to detect a hot bearing. Weather anomalies, miscalibrated scanners, and voltage spikes can all cause false alarms. Extensive braking will also heat up the wheels and bearings.

Modern roller bearings typically run at 40°F (4°C) above the ambient temperature in winter and 60°F (15°C) above the ambient temperature in summer. Continuous operation above 200°F (93°C) causes premature aging of the rubber seals and oxidation of the lubricant. At 240°F (115°C) grease breaks down and the elastomer seals begin to degrade. One major bearing manufacturer suggested 195°F (90°C) as a hot box alarm setting; another recommended 180 to 195°F (82 to 90°C).

A slow, incremental warm-up from gradual degradation has a good chance of detection. Bearing failure from massive metal-to-metal rubbing and rapid heat-up can occur in between detectors.

The very latest technology is acoustic detectors. Microphones listen for noisy bearings, and computer algorithms look for sound patterns that indicate impending bearing failures. Acoustic detectors work with fewer false alarms and can even detect many failures before heat-up.

Hot Box Detector Stops Passenger Train

On July 15, 2003, a 12-car commuter train with 1,200 passengers set off a hot box detector in New Jersey. An Amtrak dispatcher received the alarm and immediately radioed the commuter train crew to stop and inspect the train. The conductor located the correct bearing and used his temperature-sensitive crayon—but on the wrong spot. The conductor incorrectly concluded that a false alarm had occurred.

Thirty miles (48 km) later, a 600-lb (272-kg) wheel broke off the axle at 69 mph (111 km/h). The car swayed as the broken axle banged loudly on the undercarriage. A passenger recalled that "The train began to jerk, and it felt like somebody was throwing things, real hard metal, at the bottom of the car."[7] About 1 mile (1.6 kg) later, 2 cars derailed upright. Unlike what happened in the 1943 catastrophe in Philadelphia, the undercarriage did not catch on the tracks nor did a car catapult up and into solid concrete. Fortunately, the terrified passengers only had 13 minor injuries.

Derailments caused by bearing failures are extremely rare on modern passenger trains. This train struck debris that damaged the bearing seal.

BEARING FAILURE TRIGGERS CHAIN OF EVENTS

One bearing failure led to the evacuation of more than 220,000 people and the development of many new safety procedures.

The Canadian Pacific train left Windsor, Ontario, just before midnight on November 10, 1979. The train stopped 90 minutes later to pick up tank cars loaded with propane, chlorine, styrene, and toluene. The 106-car train with 3 locomotives left the train yard around six o'clock p.m., traveling east to London, Ontario.

About 89 miles (143 km) from London, residents of Milton later reported seeing smoke and sparks coming from the middle of the train. Twenty miles (32 km) farther, a hot box burned off an axle. At 50 mph (80 km/h) the dangling undercarriage of the 33rd car dragged on the track almost 2 miles (3.2 km) before derailing the train at 11:53 p.m. in Mississauga, a suburb of Toronto. Twenty-four cars derailed, including 19 cars carrying dangerous commodities. At that time there was no hot box detector on this track.

Meanwhile, a husband and wife, the Dabors, witnessed the derailment in their car at a road crossing. The Dabors noticed the dragging undercarriage and sparking without quite knowing what they were looking at. Pieces of gravel were bouncing off their car, the train cars were swaying, and the train tilted toward them. The Dabors made a hasty retreat and promptly drove into a ditch. A propane tank erupted into a fireball, and the Dabors ran for their lives. "The heat was really intense . . . I thought my hair was on fire." They ran about 0.75 mile (1.2 km). The fire seemed to be subsiding, and the Dabors decided to return to their car. A policeman stopped them and told them to run. They did, just ahead of a series of propane explosions.

Railroad procedures call for the train crew to decouple the first car and evacuate their locomotive to a safe location. The 51-year-old engineer asked his 27-year-old son-in-law trainman "if he wanted to try to free the other tankers—you don't have to but if you want to have a crack at it you can." The trainman said "Okay" and, in a fit of bravery bordering on lunacy, raced toward the fire to find the first derailed car, the 33rd. The first 32 cars were safely evacuated, including several cars filled with propane. The trainman most likely prevented additional propane explosions.

Within minutes, the large fire attracted emergency personnel. Firemen were connecting hoses, policemen were setting up roadblocks. The first order of business was to figure out what was on the train. The train manifest was unreadable. While emergency personnel were waiting for answers from Canadian Pacific, the explosions began.

At 12:10 a.m. Sunday morning, the first of three explosions knocked the Dabors, and everyone else nearby, down and spewed fire and large chunks of metal on the surrounding area. Between 5 and 10 minutes later, the second explosion occurred, followed by a third and final explosion 5 minutes later. A propane car was hurled into the air and landed more than 700 yards (0.64 km) away. The blasts were heard 30 miles (48 km) away and rattled windows for a 7-mile (11-km) radius.

BOILING PROPANE

Just as the boiling point of water increases with increasing pressure (and decreases with decreasing pressure), so does propane. Table 8.2 lists boiling temperatures for water and propane at various pressures.

Changing pressure changes the boiling temperature. This can be explained by considering the gas molecules as individual spheres in constant motion from their heat energy. Using this "kinetic theory" of gases, pressure comes about from zillions of tiny impacts of individual air molecules on a surface.

TABLE 8.2
Variation of Boiling Point of Water and Propane
with Pressure

Pressure (psia)	Boiling Temperature (°F)	
	Water	Propane
6	170	−77
14.7	212	−44
90	320	49
130	356	73

Note: There are two official pressures: psia (absolute) and psig (gage). When people refer to psi they usually mean psig. Psig is the amount of pressure greater than normal atmospheric pressure, whereas psia refers to the pressure relative to a perfect vacuum. 14.7 psia = 0 psig. 10 psig = 14.7 psia + 10 psi = 24.7 psia.

If liquid water partially fills a container, some of the liquid evaporates into the vapor space. The "jumping" of individual liquid molecules into the vapor space (i.e., evaporation) is resisted by pressure created by air molecules pounding on the liquid-air interface with their zillions of tiny impacts. If the pressure is raised by adding more air molecules into the container, there are more impacts on the liquid. Boiling becomes more difficult and requires more heat energy. The boiling temperature has been raised. If the air pressure is lowered by removing air molecules, the number of impacts at the interface decreases and the boiling tempera-ture decreases. (For the same reason water boils at a lower temperature on top of a mountain.)

Propane normally boils at $-44°F$ ($-42°C$). If the pressure is raised to 130 lbs/in^2 (896 kPa), the boiling pressure is raised to 73°F (22°C). Gas-eous propane squeezed into a liquid (by pressurizing to 130 psi) is about 220 times denser, allowing for more efficient transportation.

When a tank car filled with pressurized propane ruptures during a derailment, the pressure is suddenly released. The liquid propane in the tank car boils and expands so rapidly that it can explode. The rapidly expanding vapor results in a boiling liquid expanding vapor explosion, or BLEVE. The container is usually destroyed in a burst of flying shrapnel. A BLEVE does not require ignition and can even occur with nonflam-mable vapors such as steam. A propane BLEVE is particularly danger-ous. The propane can BLEVE and the resulting propane vapor cloud can ignite, creating a second explosion.

At 1:30 a.m. the contents of the tank cars were finally identified. Everyone's worst fears were realized. One car had 90 tons (90 MT) of chlorine, widely used to purify drinking water.

Chlorine, made famous on the battlefields of World War I, is toxic to humans. A propane tank car (in continuous danger of exploding) was next to a leaking chlorine tanker. An evacuation was ordered in the im-mediate neighborhood. With shifting winds and a nearly out of control fire, the evacuation was expanded 13 times over the next 6 hours. Even-tually, 220,000 people were evacuated.

When pressurized propane vents from a safety valve or ruptured tank and ignites, standard protocol calls for allowing the fire to burn itself out. If the burning propane jet is extinguished, the leaking propane can form

Steam BLEVE

In 1935, a French 69,000-lb (31,300-kg) boiler tore a 17-square-foot (5-square-m) hole in the side of a locomotive before flying 272 feet (83 m) and bouncing 3 times for distances of 72, 112, and 56 feet (22, 34, and 17 m). The boiler contained 14,000 lbs (6,350 kg) of pressurized liquid water heated to 392°F (200°C), squeezed to a pressure of 227 lbs/inch^2 (1,565 kPa). One way to think of all the energy released during a boiler explosion is to consider the energy required to heat up the pressurized water. (The pressure prevents the water from vaporizing into steam.) If 1 calorie (4.2 J) of heat energy raises the temperature of 1 g of water 1°C, it will take 1.136 billion calories to raise 14,000 lbs of liquid water up to 392°F (200°C). This amount of energy is equivalent to 1.32 million watt hours, or 3.5 billion foot lbs—enough energy (if focused) to lift the 69,000-lb boiler more than 50,000 feet (15.2 km) or to accelerate the boiler to a speed of 1,800 ft/sec (548 m/s).

The Union Pacific's Big Boy, one of the largest and most powerful locomotives ever made (rated at 6,290 horsepower [4,690 kW] at 41 mph [66 km/h]) weighed 772,200 lbs (350,300 kg). At full power the steam-powered Big Boy consumed 22 tons (20 MT) of coal and 100,000 lbs (45,360 kg) of water per hour.

a large vapor cloud with great potential for ignition. The propane tank with the burning vapor jet must be cooled with water, otherwise it will overheat and explode. It took almost 40 hours for all the fires to burn themselves out.

Amazingly, there were no fatalities. Chlorine, about 2.5 times heavier than air, will normally hover on the ground in a thick cloud. The Ontario Environmental Ministry estimated that 70% of the chlorine gas was carried up to 4,000 feet (1.2 km) by the updraft created by the propane explosions. It then fell back to earth in minute, harmless quantities over a 60-square-mile (155-square-km) area. The updraft is believed to have saved potentially thousands of lives. More than a mile (1.6 km) south of the explosions, residents experienced sore eyes and a sickening lump in their throats.

THE AFTERMATH OF THE MISSISSAUGA DERAILMENT

The ongoing push to replace the old-style journal-bearings with roller bearings and install hot box detectors was accelerated. "Gateway" hot box detectors were located on all tracks entering major metropolitan centers.

Today railroad companies clearly state that they would not ship hazmat if not required by law. The income does not justify the potentially open-ended liability from potential disasters. These laws date back more than 100 years and are known as "common carrier obligations."[8] Back then business would buy up all rail shipping to exclude competition. Regulators, aware that shipping by rail is far safer than by trucks, are reluctant to alter hazmat shipping patterns.[9] (Incidentally, truckers, barges, ships, and airlines can all refuse hazmat business.)

In addition to many system-wide railroad safety improvements (many described in this book), numerous improvements specific to hazardous cargo have been enacted. For example, railroads are banned from hauling dangerous cargo (i.e., chlorine) and flammable liquids on the same train.

The U.S. Department of Transportation publishes the Emergency Response Guidebook, intended to be placed in every fire and police vehicle. The American Chemical Council began providing 24-hour phone support. Most of this activity began in the early 1980s, as did the formation of most fire department hazmat teams and hazmat training for railroad workers.[10]

IMPROVED HAZMAT TANK CAR SAFETY

Head punctures once caused most tank car ruptures during accidents. After a train derailed, one car's coupling would puncture the head of another. Ordinary freight car couplings slide apart vertically. Today all hazmat tank cars are required to have reinforced couplings that resist vertical forces of at least 200,000 lbs (90,700 kg). If the cars remain attached, they are also less likely to crash into each other after derailing.

Today hazmat cars have head shields to resist puncture. The tank-head puncture-resistant system must successfully confront a coupling impact from a 263,000-lb (119,300-kg) car traveling at 18 mph (29 km/h).

As a propane tank absorbs heat from an external fire, the internal tank pressure increases. Ideally, the excess pressure is vented through a safety valve designed to open and limit pressure in just such an event. If the safety valve cannot keep up with the pressure increase, the tank can rupture or explode in a BLEVE. In both cases the blast of propane most likely ignites.[11]

In a fire steel loses 90% of its strength by the time its temperature reaches 1,800°F (982°C). Because the ability to absorb heat increases with the amount of mass available to absorb the heat, the vapor space of the tank car heats up faster and is the most likely to rupture. Since liquid propane is about 220 times more dense than propane vapor, the steel exposed to liquid propane remains cooler. For this reason firemen are trained to aim their hoses at the vapor space on the top of a tank car (or the side if the car has overturned).

Today pressurized hazmat tank cars must also have thermal barriers that maintain the steel's temperature below 800°F (427°C) after exposure to a pool fire (a burning pool of fuel) for 100 minutes and a torch fire for 30 minutes.

BETWEEN 1965 AND 1980, there were 2.2 million tank car shipments of chlorine; 788 were involved in accidents and there were 11 instances of catastrophic release, 4 of which resulted in fatalities. Since 1980, the overall hazardous materials accident rate has been reduced by 90%, and by 49% since 1990. Today railroads are required to evaluate 27 risk factors (amount shipped, population density, grade crossings, numerous track safety features, etc.) to select the safest and most secure route for shipping hazmat material.

Although greatly improved, tank cars cannot resist all crash forces, as illustrated by a February 21, 2003, burned-off axle derailment with modern roller bearings. Train 410, led by 2 locomotives pulling 77 cars, was more than 6,000 feet (1,828 m) long. A hot box detector at milepost 107.7 measured a bearing 38°F (3°C) above ambient temperature, but still within the normal range. Twenty-five miles (65 km) down the track, at milepost 82.1, the hot box detector sounded an alarm, initially a tone. When the mile-long train finished passing the detector, an automated voice advised the crew that they should stop to inspect a hot bearing on the 122nd axle. The temperature recorded was 250°F (121°C) above the ambient temperature, the highest possible reading.

Before the train could stop, 21 cars derailed at 42 mph (67 km/h), including 7 filled with pressurized LPG (liquefied petroleum gas).[12] Marks on the rail showed that one wheel of the 27th car came off the tracks first at milepost 81.5 at 5:40 a.m. At milepost 80.9 the 27th car derailed and continued east until guided by a track switch onto a siding at milepost

80.5. On the siding the 27th car struck a parked locomotive and exploded. A second tank car also derailed and exploded. Within 15 minutes 2 other derailed tank cars exploded; another exploded 4 hours later. The final car exploded at noon. The contents of all 7 loaded LPG tank cars (almost 900,000 lbs [408,200 kg]) were consumed in the fire, which burned for 3 days.

As a precaution, 300 residents were evacuated from this sparsely populated part of Ontario, Canada. There were no damages or injuries to civilians. Crew members were hospitalized with burns.

Approximately 2 minutes passed between the first warning from the detector and when the emergency brakes were applied automatically during the derailment. After braking, the 7,000-ton (7,000-MT) train continued for an additional 47 seconds before coasting to a stop.

RESEARCH CONTINUES TO IMPROVE crashworthy designs of tank cars. Computer models predict that colliding cars strike each other at half their derailment speed. For that reason trains carrying hazardous materials are limited to a speed of 50 mph (80 km/h) or less. Long-term goals are to design tank cars that can resist car-to-car impacts up to 30 mph (48 km/h) on the head and 25 mph (40 km/h) on the side. This represents a 10- to 12-fold improvement of crash energy absorption for head impact and about a 6-fold increase for side impact.[13] These goals are currently beyond the capability of any existing tank car designed with steel plate. New research hopes to develop multilayer tank car shells better able to absorb crash energy. Various steel sandwich designs (similar to corrugated cardboard) and crush zones are being studied.

ROLLER BEARINGS

Compared to a sliding block, a marble of the same weight easily rolls down the slightest incline. As demonstrated in ancient times by rolling a large stone block on a series of roller elements—logs—rolling is simply easier than sliding. As early as AD 40, ball bearings were used in turntables underneath statues. Leonardo da Vinci designed a variety of bearings very similar to modern designs. By the end of the eighteenth century, individual inventors were using ball bearings in windmills, water turbines, and carriages. The basic concept involves balls rolling between

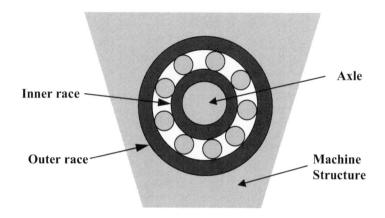

FIGURE 8.10. Assembled bearing. The inner race is fixed on the axle, the outer race is fixed to the machine.

inner and outer cylinders, or races. A separator, or "cage," keeps the balls from rubbing.

The inner cylinder is attached to a rotating axle, usually by an interference fit. This means the inner race has a smaller inside diameter that is forced or pressed onto the larger outside diameter of the axle. The outer race is then pressed onto the machine (see Figure 8.10).

Until better-quality inexpensive steel became available, ball bearings were made with cast iron balls. Cast iron was brittle and would easily split under heavy loads.

The contact area and load-carrying capacity of a ball bearing can be increased by replacing the balls with cylinders. A bearing using cylindrical rolling elements is called a roller bearing. Roller bearings, however, cannot support loading parallel to the axle.

To solve this problem Henry Timken invented the tapered roller bearing in 1898 for horse-drawn carriages. He also started the highly successful Timken Company. Today the company remains a major supplier of tapered roller bearings for all applications, including railroads. Although tapered roller bearings were successfully used in automobiles for years (and appeared occasionally on lightly loaded passenger train cars in the 1920s), the railroad companies remained skeptical about roller bearings on heavy freight trains.

Eventually, the Timken Company gave up trying to convince the railroads of its better mousetrap and adapted a locomotive for roller bearing use. In a 1930 publicity stunt 3 women demonstrated the low starting force of roller bearings by pulling a 650,000-lb (294,800-kg) locomotive in high heel shoes! The force to start the locomotive with roller bearings was an estimated 200 lbs (890 N) compared to 6,000 lbs (26.7 kN) for plain journal-bearings.

An occasional wheel with a flat spot and/or misaligned track causes severe impacts that are hard to quantify and design for. Railroads were still concerned about concentrating that much impact loading onto the tiny contact area of the rollers. Also, roller bearings cost 20 times more than plain journal-bearings. Timken finally sold 10 locomotives equipped with roller bearings in 1934. Finally, in 1947, 1,000 freight cars were purchased with tapered roller bearings. New technology always appeared in passenger train cars first before finally reappearing in the nation's then 1.7 million freight cars. By 1968, 25% of freight cars had roller bearings, 50% by 1975, and about 90% by 1986. Roller bearings became required on new freight cars in 1968 and finally on all cars in 1994. Bearing-related derailments rapidly declined with widespread use of the more reliable roller bearing, as shown in Figure 8.11.

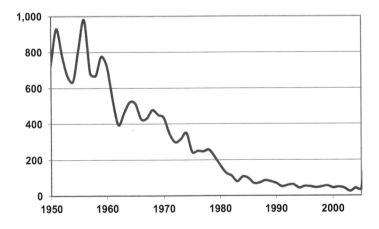

FIGURE 8.11. Derailments due to axle burn-off from bearing failures.

FREIGHT CAR ROLLER BEARINGS

A tapered roller bearing consists of an inner cylinder commonly called a "cone," an outer cylinder known as a "cup," and tapered rollers. The rollers, inner and outer cylinders, carry the load. To increase load-carrying capacity, two rows of cones and rollers are used on railroad axles (Figure 8.12). One common configuration is 23 rollers in each row (46 total) with diameters of 0.84235 inch (2.13957 cm). Not shown are

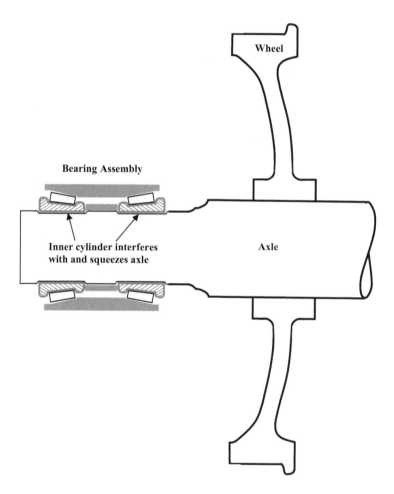

FIGURE 8.12. Two inner cylinders, a spacer in between, two sets of rollers, and one outer cylinder. Also shown (greatly exaggerated) is the interference fit between the inner cylinder and axle.

FIGURE 8.13. Wheel and bearing attached to an axle.

the cages that maintain proper spacing of the rollers, the end cap, and the seal that clamps the bearing on the axle.

The cone is manufactured to have an interference fit with the axle of 0.0025–0.0045 inch (0.0635–0.1143 mm) (i.e., the cone is smaller than the axle and pressed onto the axle). Railroad bearings are hydraulically pressed onto the shaft. A typical pressing force is 100,000 lbs (445 kN). The bearing is clamped against a radius on the axle with an end cap bolted on the end of the axle. A wheel and bearing attached to an axle is shown in Figure 8.13.

The axle, wheel, and inner race of the bearing all spin as one attached unit. The outer cylinder of the bearing is kept from spinning by the bearing adapter that fits in the side frame, as shown in Figures 8.14 and 8.15.

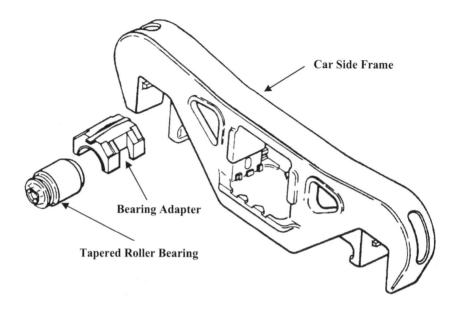

Car Side Frame

Bearing Adapter

Tapered Roller Bearing

FIGURE 8.14. Bearing, bearing adapter, and car side frame.
From FRA/ORD-80/43.

FIGURE 8.15. Two wheels, bearings, and the car side frame. Also shown is the
center bolster sitting on springs.

The car actually sits on a spring-mounted center bolster. The springs provide some flexibility for rail and wheel anomalies and during cornering.

ROLLER BEARINGS: ADVANTAGES AND DISADVANTAGES

The biggest advantage to using roller bearings is reduced delays caused by hot bearings. When a bearing overheats, the engineer must drive the train at 5 mph (8 km/h) (with his or her head hanging out the window looking for a smoking bearing) until a siding can be located to remove the faulty car. Fewer derailments and reduced starting force are also major benefits.

As opposed to journal-bearings that need constant attention, roller bearings are permanently sealed and designed to last the life of the wheel without any further lubrication or other maintenance. Roller bearings are expensive because the rollers, cups, and cones are ground to an incredible (and costly) tolerance of 50 millionths of an inch (0.00127 mm). "Super finishing" is done to all surfaces to minimize microscopic contact and wear. Failure, when it does occur, occurs faster than in a plain journal-bearing because of the smaller contact area.

Operating procedures, and sometimes even the layout of train yards, had to be changed after the introduction of roller bearings. In a yard a pushed freight car is expected to roll to a stop after a predictable distance. With roller bearings, cars rolled faster and farther. Also, freight cars sitting on an incline can more easily roll away, becoming dangerous "runaways" (see Chapter 9).

ROLLER BEARING FAILURE

Incremental design improvements and hot box detection continue to improve reliability. Failures continue, albeit greatly reduced.

Roller bearings are rated for a specific load and number of operating hours or rotations. If these limits are exceeded, the bearing will eventually fail from wear and/or rolling contact fatigue very similar to how rails (see Chapter 11) and wheels fail. Subsurface metal fatigue grows until a surface pit, or "spall," appears with small particles flaking off. The inner cylinders, outer cylinders, and rolling elements are all susceptible to spalling.

As the wear particles accumulate, the lubricant will eventually break down and fail to keep the rolling elements separated by a thin layer of grease. The lubricant can also break down at elevated temperatures and from accumulated water. Degraded grease generates more heat, which further accelerates grease breakdown.

Based on "pristine" laboratory testing, 90% of railroad bearings are expected to last 1 million miles (1.6 million km) before spalling or wear occurs. Actual freight cars sway and run on wheels and tracks with imperfections. These "real-world" conditions reduce the expected laboratory life to perhaps 500,000 to 600,000 miles (800,000 to 965,000 km). Ideally, the wheels wear out or fail first, maybe after a few hundred thousand miles (km). At that time the bearings are serviced.

Bearing failure is a common problem in all machinery. There are additional failure modes unique to heavily loaded railroad cars. The wheels can develop flat spots from skidding during braking. Impact loads can be more than four times the normal load (see "Wheel Impact Loads" in Chapter 11). Another problem is flexing of the axle. A modern freight car can weigh up to 286,000 lbs (130,000 kg). The axle bends as shown (exaggerated) in Figure 8.16.

The bending results in elongation of the top surface of the axle and compression on the bottom. The elongation and the compression reverse with each rotation. This cyclic elongation/compression plays out as constant rubbing between the inside diameter of the bearing and the outside diameter of the axle. The bearing and the axle can rub 100 million times between one scheduled routine maintenance and the next.

The rubbing degrades the interference fit between the bearing and axle and may cause the bearing to slip on the axle. Initially, this slip is very small, maybe 1 rotation after 1,000 miles (1,600 km). The slippage slowly causes more wear and increased slippage. The harder bearing material always wears the softer axle. The diameter of the axle slowly becomes smaller and visibly grooves under the bearing inner cylinders. Eventually, the inside diameter of the bearing is running freely on the axle with massive rubbing friction and heat generation.

One potential "fix" is to make the axles stiffer to reduce the flexing shown in Figure 8.16. Unfortunately, replacing the existing 4 × 1.3 mil-

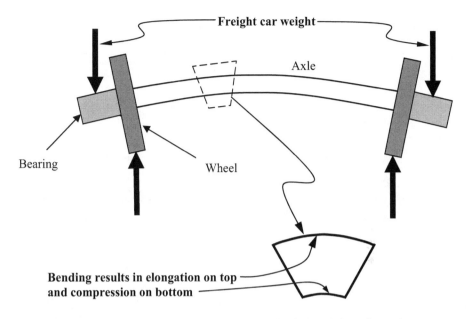

FIGURE 8.16. The loads on the bearings and wheels bend the axle, as shown (exaggerated). Also shown is a section with elongation on the top and compression on the bottom.

lion freight cars = 5.2 million axles is not an option, and the bearing designer must live with the existing configuration and axle flex.

Burned-off axles are almost completely unheard of in the modern passenger era. Passenger train cars do not have a legacy problem. With only 50,000 passenger cars, the fleet has been replaced many times over with axles that flex less. Also, the people on the train can often smell a smoking bearing (or hear a noisy bearing). Most high-speed passenger trains have a temperature sensor on each bearing.

Today about 4% of all freight bearings (more than 500,000) are replaced each year in the United States, 99.99% proactively before any derailment. A similar number of wheels are also replaced each year.

In one survey 12% of replaced freight bearings were destroyed by heat. Because the parts were partially melted, no cause could be determined. Twenty-five percent were replaced for nonbearing problems (damaged during derailing, severe wheel impacts, etc.). The rest were replaced after inspection (with visible damage) or they set off an alarm; 23% of them had loose components, 13% spalling, 4% lubrication breakdown, 3% damaged seals, and the rest had corrosion from water.

BEARINGS FAILED ON PURPOSE

Bearing failures are common in industry and even in everyday life. Burned-off axle failures, however, are almost exclusive to railroad freight cars. In other machinery the bad bearing becomes noisy and hot. The problem is noticed by operators, the bearing is replaced, and no physical damage is done to attached machinery. Because railroad bearings are heavily loaded and people can be a mile (1.6 km) away, metal-to-metal rubbing occurs until the axle burns off. Forensic examination usually cannot identify the initial failure mechanism because the end result is a pile of partially melted metal.

For this reason there have been several test programs to study the problem with actual trains. One test studied the gradual development of inner cylinder slip on the axle. In this test 20 used bearings removed for overheating were installed on axles worn from bearing slip. Recall that the design interference fit is 0.0025 to 0.0045 inch (0.0635 to 0.1143 mm). The used and worn bearings had interference fits less than 0.0015 inch (0.0381 mm).

The 2 worst bearings slipped continuously up to 1.6%. This means the bearing inner cylinder slipped 1.6 rotations every 100 wheel rotations. (The other bearings with better interference fits only slipped 2 to 7 rotations during 5,000 miles [8,046 km] of testing.) The 2 worst bearings were recorded at 258°F (126°C) and 486°F (252°C) when running at 60 and 50 mph (97 and 80 km/h). The temperatures for the remaining used bearings tested at 70 mph (113 km/h) ranged from 65 to 140°F (18 to 60°C) above the ambient temperature.

The data showed that bearing slip on the axle is a very gradual process related to the quality of the interference fit and the clamping force. As the interference fit and clamping force slowly degrade, slip accelerates. Initially, minor slip generates little frictional heat, but the continuous wear of the axle eventually leads to misalignment of the roller elements, as shown in Figure 8.17.

The misalignment overloads the roller elements. As the bearing begins to heat up, it thermally expands. The bearing, with a now larger

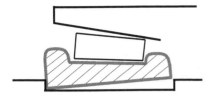

FIGURE 8.17. Wear on an axle results in misalignment and premature failure of roller elements.

diameter, loosens and slips more, creating additional heat. Overloaded rollers eventually spall and wear to a smaller size. In time the undersized roller wedges into the separator cage, jams, stops rotating, and generates massive heat. The rollers can also degrade from ordinary wear, contaminated grease, faulty installation, and/or wheel-induced impacts.

A different test tried to measure bearing temperature during seizing. The bearings on an unloaded freight car normally operated at 30–40°F (16–22°C) above the ambient temperature at 30 mph (48 km/h). This increased to 45–55°F (25–30°C) at 50 mph (80 km/h).

The grease was removed to purposely seize the bearings. Grease removal mimics what actually happens in an overheated bearing. The grease will eventually cook off or become significantly degraded with wear particles. After 15 minutes at 30 mph (48 km/h), the temperature began to increase and the bearing seized after 18 minutes. The thermocouple read 375°F (190°C) before it was wiped off during seizure.

The tests were repeated with another degreased bearing in July (95°F [35°C] ambient temperature) and a car loaded to 232,500 lbs (105,500 kg) gross. At 30 mph (48 km/m) and 50 mph (80 km/h), the bearing heated up to 160°F (71°C) and 200°F (93°C).

After 1 hour at 30 mph (48 km/h), the bearing showed no signs of overheating. The speed was increased to 60 mph (97 km/h), and the bearing began a near uniform temperature rise of 3–4°F (1.6–2.8°C) per minute for 45 minutes. When the temperature reached 360°F (182°C), it was decided (out of concern for the safety of the train crew) to reduce the speed to 30 mph (48 km/h) to induce failure at a lower speed. The temperature cooled, and after 30 minutes there was no temperature increase. The speed was increased to 60 mph (97 km/h). After 10 minutes, the temperature began to rapidly increase. At consecutive 15-second intervals the temperatures recorded were 320°F, 415°F, 1,195°F, and 1,531°F (160°C, 213°C, 646°C, and 833°C). Somewhere between 415°F

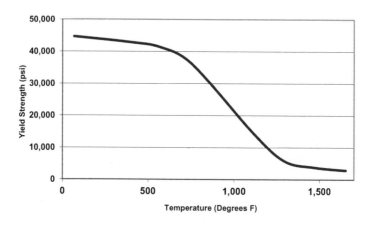

FIGURE 8.18. Strength of axle steel versus temperature. *Adapted from Wang's Ph.D. dissertation.*

and 1,195°F, the bearing seized. As soon as the bearing seized, the speed was reduced to 30 mph (48 km/h). After only a few seconds, the bearing broke free, began to spin, and cooled with consecutive 15-second readings of 1,531°F, 851°F, 703°F, and 647°F (833°C, 455°C, 373°C, and 342°C). The bearing continued to cool. When the residual grease degrades with wear particles, it has the consistency of fudge. Once the degraded grease heats up, liquefies, and rewets the bearing components, rotation can resume and the bearing cools.

Trying to re-seize the bearing, the speed was increased in increments to 45, 50, and 55 mph (72, 80, and 89 km/h). The bearing continued to rotate. By this time it was decided to end the test because of overtime costs for the train crew, fire department, and security personnel.

The rapid heat-up also degrades the strength of the axle (Figure 8.18). The extreme heat and rotation neck the axle down to a conical stub until it fractures.

RAPID BURN-OFF

Because of rapid heat-up from metal-to-metal rubbing, the axle can burn off quickly. Of 47 burned-off axle derailments surveyed in Canada between 1992 and 1996, axle failure occurred on average 11.5 miles (18.5 km) after a hot box detector. And 22 of the derailments occurred just 5 miles (8 km) past the detector.

One such rapid burned-off axle derailment occurred on October 19, 2003, in Canada. After passing hot box detectors at mileposts 140.7 and 123.0, a hot bearing warning was broadcast to the train crew at milepost 98.5. The radio, however, had been inadvertently set to the wrong channel. (Each crew member had heard only two hot box detector alarms in the previous year.)

Pieces of bearing were found at milepost 97.4 and the train derailed at milepost 96.41. The 6,061-foot (1.85-km)-long 45-car train continued until it came apart and coasted to a stop at milepost 86. Two cars derailed upright with minimal damage. The train had been traveling between 35 and 45 mph (56 and 72 km/h).

9

Gravity

It's the Law

Unlike a sudden, violent derailment or collision, a runaway train is our only slow-motion disaster—at least at first. Everyone involved slowly becomes aware of the problem before their imagination runs away with what happens next. Often a panic decision is made: "ride it out" or jump with uncertain safety. The runaway train doesn't have to end in a high-speed collision or derailment, but of course the famous ones do.

THE ARMAGH DISASTER

On June 12, 1889, the Armagh, Ireland, Sunday school began their annual outing to a seaside resort. A marching band led a parade to the train station. Children paraded with flags and banners. The whole town turned out to see them off. With 941 people, mostly children, crowding into 14 cars, 2 coaches were added at the last minute. In defiance of modern safety practice, the doors were locked to keep the children in and fare-dodgers out. It's believed nonpaying passengers added to the load.

The engineer expected a more powerful locomotive but still believed he could make it up the first 2.5-mile (4-km)-long hill. The train stalled just short of the crest. The crew decided to uncouple the first 4 cars, take them to the next station, and come back later for the remaining 10 cars. With no brakes holding the remaining cars, the conductor wedged stones under the wheels of the last two cars. Approximately 600 passengers were left waiting in 10 cars held by rocks.

When the train stalled, the locomotive's steam piston stopped at the center stroke position. The center stroke position is a dead spot in the cycle with little power. When the locomotive restarted, it lurched backward and shifted the rocks.

The 10 cars started slowly rolling backward. The conductor unsuccessfully tried to wedge more rocks. As the passengers gradually realized their plight, they started passing children out the windows before jumping after them.

Meanwhile, a second train left Armagh just 10 minutes behind the runaway train. When the second train reached the foot of the hill, the crew saw the 10 cars moving backward toward them. The train crew managed to slow their train to 5 mph (8 km/h) before jumping and yelling at their passengers to do the same. No injuries were reported on the second train. The runaway train was traveling backward at an estimated 50–60 mph (80–97 km/h). The collision destroyed the last 3 cars of the church train and killed 78 passengers.

In a determined bid to repeat history, a train got stuck climbing a hill in Algeria in 1981. A second locomotive was sent to help. Instead of simply connecting the second locomotive to the stopped train, for some unexplained reason the trainmen uncoupled the original locomotive before properly securing the passenger cars. The train rolled backward into a freight train, killing more than 130. At least there were no rocks involved.

But there were rocks involved in Mozambique in 2002. A passenger train carrying more than 1,000 passengers was parked on a hill with 4 ill-advised large rocks under the wheels. The train rolled backward into a freight train, killing 196. In that same year a passenger train in Tanzania, also carrying about 1,000 passengers, rolled backward (from an undisclosed mechanical failure) for 25 minutes, reaching 125 mph (201 km/h) before striking a freight train, killing 281.

In 2004, a Washington Metro train ran away backward and struck a stopped train at 34 mph (55 km/h). One car was crushed nearly 34 feet

(10 m). Because both trains were nearly empty, there were, fortunately, no fatalities. Apparently, the hand lever got stuck between two brake settings, it failed to engage, and the operator did not apply the emergency brakes.

Runaway trains still occur today if the train is too heavy, is moving too fast, or is left unsecured on a hill. Brake failures, discussed in the following chapter, can also occur.

CARS LEFT UNSECURED

This twenty-first-century story starts with confusion in a train yard and ends with the excitement of routing 31 runaway cars headed toward downtown Los Angeles at 95 mph (153 km/h). The immediate task was avoiding collision with a nearby passenger train and a freight train transporting explosive liquefied petroleum gas.

On June 20, 2003, a freight train with 3 locomotives and 31 cars stopped at Montclair Yard to change crews and drop off cars. As the train entered the yard, the dispatcher instructed the train crew to enter a siding and contact the yard's switching crew.

Operating rules require that all cars not attached to a locomotive be secured with hand brakes. Hand brakes, an artifact from the nineteenth century, consist of a hand wheel on each car that activates the brake shoes through a series of linkages and levers. The switching crew told the incoming train crew to skip the hand brakes because they planned on immediately attaching their locomotive to the 31 cars. Although this violates regulations, this procedure was commonly done at this yard to save time. The arriving crew uncoupled the 31 cars and left.

The detached 31 cars were secured by their emergency brakes. For safety purposes emergency brakes automatically activate when the airline is disconnected anywhere in the train. Emergency brakes use air pressure in the brake cylinder to press the brake shoe against the wheel. To move a car secured in this fashion the air pressure in the brake cylinder must be released manually.

The switching crew began releasing the emergency airbrakes on the 31 detached cars. Two trainmen started in the middle and worked their way to opposite ends of the train. They completed their task without noticing that there was no locomotive attached to either end of the

3,380-ton (3,066-MT), 2,281-foot (695-m)-long string of cars, and both walked away to attend to other duties.

Shortly after, the crew noticed the 31 cars heading west out of the yard toward Los Angeles. Although the train yard's grade was not recorded, a little bit of inattentiveness will lead to a runaway train on even a modest 0.5% grade—just a 6-inch (15-cm) drop every 100 feet (30 m). A fully loaded train rolling with gravity would take just 77 seconds and 427 feet (130 m) to reach 8 mph (13 km/h).[1] In another 41 seconds the train is moving 12 mph (19 km/h). Adjusting for realistic values of friction and rolling resistance that slow the train slightly, the distances increase by 10% to 15%. Everything occurs faster on a steeper grade.

The switching engineer notified the yard dispatcher around 11:33 a.m. The runaway cars entered the main track 1 minute later. The incoming crew wanted to chase after the runaway with their locomotives. This option was briefly considered and rejected. Chasing locomotives can collide with the runaway if it starts to roll backward. In fact, switches were thrown to prevent the cars from coming back into the yard and causing damage or injury. Because of confusion (about whether the cars would stop, reverse direction, or accelerate), dispatchers did not notify local authorities immediately.

Meanwhile, some Pomona City police officers, stopped at a crossing about 4 miles (6.4 km) from the yard, noticed the 31 cars passing without a locomotive attached. The police called the 800 number posted at the crossing and immediately reached the Union Pacific Response Management Communications Center in St. Louis, Missouri. The police estimated the runaway's speed at 60 mph (97 km/h).

A number of switching options were considered, including onto various sidings along the way or onto a different main track. (In this location all switches are remotely controlled by the dispatcher.) Mindful of nearby freight and passenger trains, they had to select the best location to derail the out-of-control train headed for downtown Los Angeles.

After 26 minutes and 28 miles (45 km), the runaway finally derailed at an estimated speed of 95 mph (153 km/h). The derailment destroyed 3 houses, damaged 5 others, and injured 13 people.

Many yards protect the main track with a derailer. A derailer is placed on the rail to prevent a runaway train from accidentally entering the main track. This yard did not use a derailer. After the accident, Union Pacific installed derailers to protect all main tracks from runaway equipment on sidings used for the storage of cars when the grade[2] exceeds 0.5%. All dispatchers were issued written procedures for runaways in their district.

THE LONGEST RUNAWAY?

The record for the longest runaway train will never be known. One leading candidate occurred on March 27, 1884. Eight loaded box cars standing on a siding in Akron, Colorado (elevation 4,714 feet [1,436 m]), were blown by a strong wind over a switch set to protect the main track from just such an event. A freight train took chase in Benkelman, Nebraska (elevation 3,000 feet [914 m]), hoping to couple onto the runaway. After a 5-mile (8-km) pursuit, the runaway was caught, but not before coasting 96 miles (154 km)!

NEWTON'S LAW, ACCELERATION, AND RUNAWAY TRAINS

Acceleration is how fast speed changes. The most familiar acceleration is that of falling objects. If air resistance is neglected, the speed gradually and uniformly increases by 32 ft/sec (9.8 m/s) for every second of free fall. The speed after 1 second is 32 ft/sec (9.8 m/s), after 2 seconds 64 ft/sec (19.6 m/s), and so forth. The acceleration of gravity increases the object's speed by 32 ft/sec every second. Gravitational acceleration is said to be 32 ft/per second per second, or 32 ft/sec^2 (9.8 m/s^2).

The average falling speed is one-half the final speed. For example, the average speed after falling for 3 seconds is the average of the initial speed (zero) and the final speed (96 ft/sec) = 48 ft/sec (14.6 m/s). The distance fallen equals the average speed × duration of falling. After 3 seconds, the distance fallen = 48 ft/sec × 3 seconds = 144 feet (43.9 m).

In general, distance = average speed × time and average speed = 0.5 (acceleration × time). Substituting the average speed into the distance equation gives distance = 0.5 × acceleration × time^2. After 4 seconds, the proverbial fallen apple will have dropped

$$0.5 \times 32 \text{ ft/sec}^2 \times 4^2 = 256 \text{ feet (78 m)}$$

TABLE 9.1
Speeds and Distances for Free Fall

Time (sec)	Maximum Speed (ft/sec)	Maximum Speed (mph)	Average Speed (ft/sec)	Total Distance Dropped (ft)
1	32	21.9	16	16
2	64	43.9	32	64
5	160	109	80	400

Table 9.1 shows the relationships between time, speed, average speed, and total distance fallen for an object falling with gravity.

A train falling off a cliff for 5 seconds falls 400 feet (122 m) and hits the ground at a speed of 109 mph (175 km/h). Ignoring friction (a good first approximation), a runaway train dropping 400 feet (122 m) in elevation on a 2% grade (or any other grade) will reach the same speed as falling off the cliff 400 feet.

Just as Galileo concluded that all objects fall at the same speed regardless of weight, all runaway trains on the same hill roll downhill with the same acceleration and reach the same speed.[3]

It turns out (through a few tricks of trigonometry) that a train rolling down any grade is being accelerated along the tracks by a percent of the train's weight equal to the grade. In other words, a train on a 2% grade is accelerated by a force (parallel to the tracks) equal to 2% of its weight and a train on a 1% grade is being accelerated by 1% of its weight. For convenience, we will consider one freight car weighing 200,000 lbs (90,700 kg).

Using Newton's Law for the 200,000-lb freight car, the car is being accelerated on a 2% grade by a force equal to 0.02 × 200,000 lbs = 4,000 lbs (17.8 kN). (The weight must be divided by the acceleration of gravity to convert weight to mass. See "Weight versus Mass" box in Chapter 4.) The formula $f = m \times a$ becomes:

$$4{,}000 \text{ lbs} = \frac{200{,}000 \text{ lbs}}{32 \text{ ft/sec}^2} \times \text{acceleration}$$

Solving the above equation, the acceleration is found to be 0.64 ft/sec² (0.196 m/s²), or 2% of the normal acceleration of gravity.

Since a 2% grade drops 2 feet (0.61 m) for every 100 feet (30.5 m) of horizontal track, it takes 200 × 100 feet, or 20,000 feet (6,096 m), of track before the track drops 400 feet (122 m). The time the train takes to travel 20,000 feet while accelerating at 2% of normal gravitational acceleration is:

$$20,000 \text{ feet} = \frac{1}{2} \times \frac{0.64 \text{ feet}}{\text{second}^2} \times \text{time}^2$$

Solving the equation, time = 250 seconds. Accelerating for 250 seconds at 0.64 ft/sec² (0.196 m/s²) gives a final velocity of 250 seconds × 0.64 ft/sec² = 160 ft/sec (equal to 109 mph [175 km/h]), the same speed if the train dropped straight down off the cliff. If all the calculations were repeated for a 1% grade, the track would be 40,000 feet (12,192 m) long before it dropped 400 feet (122 m), the force accelerating the car would be 1% of the car's weight, and the acceleration would be 1% of gravitational acceleration. It would take 500 seconds for the car to fall 400 feet on the 1% grade and the speed at the bottom of the grade would be the same 160 ft/sec.

The same gravity force works in reverse. A 4,000-lb (17.9-kN) force must be applied by a locomotive to pull the 200,000-lb (90,700-kg) car up a 2% grade.

TRAIN RESISTANCE

The force required to pull a train at constant speed on level track is known as train resistance. Train resistance depends on speed, weight, type of cars, condition of track, brand of bearings, and many other factors. Because it affects fuel consumption, train resistance has been widely studied.

Train resistance is estimated by multiplying the train's weight, velocity, and velocity squared by constants. The constants vary from study to study, and railroad companies develop their own values based on their experience. Shown in Figure 9.1 is estimated resistance for a 75-ton (68-MT) and a 100-ton (90-MT) car at various speeds.

Train resistance is normally considered to consist of the following: rolling resistance of the wheels and track, bearing resistance, rubbing between the wheel flange and track, and air resistance.

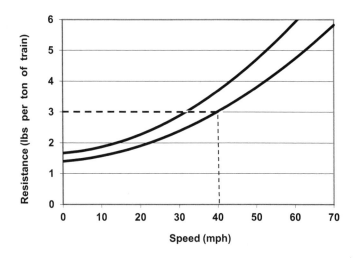

FIGURE 9.1. The two curves estimate resistance for a 75-ton (top curve) and 100-ton car. A 100-ton car traveling at 40 mph has a resistance of $100 \times 3 = 300$ lbs.

ROLLING RESISTANCE

Demonstrating the obvious advantage of rolling versus sliding, a marble rolls downhill better than a cube slides. A small force is still required, however, to maintain rolling.

Consider a wheelbarrow, with air-filled rubber tires, being pushed at constant speed. According to $f = m \times a$, the pushing force should result in constant acceleration—unless there is an equal and opposite force. The opposing force is the rolling resistance of the wheelbarrow.

Starting with the definition of work = force × distance, the tire absorbs work and energy every rotation when the tire contact point deforms into a flat spot. The rotating tire continuously requires additional work to flatten a new spot on the tire. The energy dissipated by rolling reappears as the energy or work required to push the wheelbarrow (work = rolling resistance × distance pushed). A tire with low air pressure deforms more, absorbs more energy, and is harder to push. Noticeably worse mileage occurs with underinflated tires.

The solid steel railroad wheels create a flat spot about the size of a dime. Nevertheless, this deformation represents a force (the load on the wheel) times a distance (the flattening of the wheel). The steel wheels have about one-tenth the rolling resistance that a road vehicle's pneumatic tires have—an important part of the economic advantages of railroad transportation.

Roller bearings have similar deformations and rolling resistance. The rail also deforms and absorbs energy. Rolling resistance increases as the track degrades. Loose track creates more rubbing and deformation.

AIR RESISTANCE

Air resistance occurs from air rubbing on the surfaces of the train, air impact on the front face, and separated flow in the back of the train. As the train moves through the air, there is a slight vacuum immediately behind the last car. Turbulent air rushes into the vacuum. Aerodynamicists refer to this as separated flow. The cloud of dust behind a car driving down a dirt road illustrates separated flow. It takes energy to stir up the air. Because of the many types of freight cars, a precise prediction of air resistance is difficult. Air drag increases with velocity squared and will be the dominant resistance at high speeds.

Air resistance also increases with train length. The air drag on a short train of 10 cars at 80 mph (129 km/h) is 50% of all train resistance. A longer train of 68 boxcars at 80 mph has air drag that is 79% of all resistance. One study gives the following breakdown: at 50 mph (80 km/h) a loaded freight car has about 16% of the train resistance from the bearings, 56% from the wheels, and 28% from air resistance. The same car empty is 60% aero, 22% bearing, and 18% wheels. (With less deformation of the wheels, rails, and bearings, there is less rolling resistance in an empty car. The air resistance, staying the same, makes up a bigger percentage.)

Total train resistance increases with weight. Resistance per ton decreases, however, because air resistance remains the same. Also, bouncing and rubbing on the rails reduce with increased weight.

Train resistance becomes a relatively minor adjustment to Newton's Law. For example, on a 2% grade the gravity force is 40 lbs per ton versus about 3 lbs per ton for train resistance. Newton's Law for a runaway train adjusted for train resistance becomes:

$$\text{gravity force} - \text{train resistance force} = \text{mass} \times \text{acceleration}$$

MAKING THE GRADE

Making the grade is the most important issue in railroading. The choices are up and over, around, or through. Once the railroad decides to operate between points A and B, all other layout considerations almost totally depend on minimizing the grade.

The rate of rise or grade is usually described as a percent. A 1% ascending grade means the hill rises 1 foot (0.30 m) for every 100 feet (30 m) of horizontal length—an angle of just 0.573 degree. The grade at Armagh was 1.33%.

The grade substantially controls operating costs and affects the economic viability of the railroad from day one. Just a modest 0.5% grade more than triples operating costs. The company must balance the economic penalty of borrowing heavily to build it right the first time (and risk bankruptcy with excessive debt) versus the higher operating costs and safety risks of using a more direct, but steeper route. Historically, many railroads have gone bankrupt by making these decisions poorly.

With a typical train resistance on level ground of 3 lbs per ton, a 10,000-ton (9,071-MT) train has a resistance of 3 lbs/ton × 10,000 tons = 30,000 lbs force (133 N). This train requires 30,000 lbs (133 N) of force to maintain a constant speed on level track.

A train ascending a 0.5% grade needs an additional force equal to 0.5% of the train's weight to compensate for gravity. This additional gravity force equals 0.5% of 10,000 tons, or 100,000 lbs (444 N), of additional locomotive pull. The 10,000-ton train now needs 30,000 lbs + 100,000 lbs, or 130,000 lbs (578 N), to climb the 0.5% grade. The required locomotive force for various grades is summarized in Table 9.2.

The smallest hill requires significantly more pulling force. If one locomotive can pull 30,000 lbs (133 N) and move the 100-car train, 8 locomotives are needed to pull the same train up a 1.0% grade. The 8 locomotives will require 8 times more fuel, initial capital, and maintenance costs, and, for steam locomotives, 8 times the labor cost.

Historically, many methods have been used to make the grade. The railroad might construct tunnels through solid rock, or loop the track back and forth to gradually climb the mountain. Another possibility is to

Table 9.2
Required Locomotive Force for 10,000-Ton
Train Required to Climb Various Hills

	Required Locomotive Force (lbs)
Level ground	30,000
0.5% grade	130,000
1.0% grade	230,000
2.0% grade	430,000

build more powerful locomotives or locate helper engines at particularly steep grades. For lesser traveled routes, the train would be split into two, three, or four sections, making it easier for the underpowered locomotive to make it up a steep grade.

In the 1960s, the railroads started using two, three, or four diesel locomotives controlled by one train crew. They might even use seven or eight on a particularly heavy coal train. In many cases steep grades have been abandoned or redesigned with lower grades.

The ruling grade on a given route is the most difficult hill to climb and determines how much tonnage can be hauled by a given set of locomotives. The ruling grade is not necessarily the steepest grade, as a short hill can often be overcome by the train's momentum with a running start.

Maximum Grade: 2.2%

A 2.2% maximum grade is an unofficial limit in North America.

After private investors refused to fund a railroad to California, the U.S. government stepped in. The government wanted assurances it was funding a sensible project. Civil War general George B. McClellan, an army engineer, was assigned the task of surveying U.S. railroads for best practice. McClellan cited the Baltimore & Ohio's use of 2.2% grades across the Alleghenies as a practical limit. The Pacific Railroad Act of 1862 states that the grades and curves will not exceed the grades and curves of the Baltimore & Ohio Railroad.

Most government-assisted railroad construction projects in the United States and Canada as well as many privately funded projects used similar language. The 2.2% limit was not a codified law. Railroads occasionally exceed it. Banks considered steeper grades as riskier, however, and demanded a higher rate of return.

In the United States the steepest main line grade (not an industrial spur) was the 3-mile (4.8-km)-long 4.7% Saluda grade in North Carolina. With a long history of runaways, Saluda Pass was closed in 2001 because of changing coal traffic patterns. BNSF's 3.3% Raton Pass in New Mexico is now believed to be the steepest main line grade in North America.

The famous Kicking Horse Pass, built by the Canadian Pacific Railway to cross the Continental Divide of the Canadian Rockies, illustrates many of the problems of a steep grade.

Kicking Horse Pass

In the middle of the nineteenth century, the good citizens of British Columbia were beginning to look south to their North American neighbors for economic development. They agreed to a union with eastern Canada only if connected by railroad. At many locations there were no known passes through the Canadian Rockies. Between 1871 and 1880, more than 800 men completed compass and barometer surveys.[4]

Six passes, ranging in elevations from 3,711 feet (1,131 m) to 6,663 feet (2,031 m), were located at the Continental Divide. The lowest pass is always the best. The Canadian government feared incursions from the northernmost U.S. railroads, however, and unexpectedly made the controversial decision to choose Kicking Horse Pass (5,337 feet [1,626 m] in elevation) for its southern location. Construction began in 1881.

Maximum grades of 2.2% were originally planned. At Kicking Horse Pass across the Continental Divide, the mountain river drops 1,100 feet (335 m) in 3.5 miles (5.6 km), a grade of almost 6%. To meet the 2.2% limit, the track would need to pass over several unstable boulder fields, under a glacier, and still require a 1,400-foot (427-m) tunnel through solid rock. It was decided to temporarily construct a 4-mile (6.4-km) stretch of 4.4% grade, followed by a level stretch and then another steep grade. The very first train to descend (carrying a construction crew) ran away, killing 3.

The line, opened in 1885, had safety switches installed every 2 miles (3.2 km) to catch runaway trains. The switches, manned 24 hours per day, redirected trains onto an uphill spur. Only when a train was visually verified to be traveling at a safe speed was the switch thrown to allow continued descent.

At the crest every passenger train stopped for a manual inspection of all brake equipment. Brakemen would jump off and run along the train at various sites to inspect for sliding wheels or overheated brakes. During descent, freight trains were restricted to 6 mph (9.7 km/h).

IN SPITE OF ALL THE PRECAUTIONS, runaways still occurred on Kicking Horse Pass. Besides the safety hazards, the operational delays and expenses getting up the "Big Hill" justified change. In 1909, the famous Spiral Tunnels opened and reduced the grade to a manageable 2.2%. Two locomotives, instead of four, could now be used.

The first tunnel, more than 3,200 feet (975 m) long, circles for 291 degrees, creating a bizarre visual—both ends of the same train can be seen moving in opposite directions when entering and leaving the tunnel. The second tunnel, just short of 3,000 feet (914 m), circles 217 degrees. The tunnels, still used today, required 1,000 men with 75 carloads of dynamite working almost 2 years to complete (Figure 9.2).

Descending a 2.2% grade is still a challenge with little margin for error. Runaways still occur on Kicking Horse Pass. A 112-car train with 4 locomotives weighing 14,800 tons (13,426 MT) and approximately 6,900 feet (2,103 m) long had a runaway just past the Upper Spiral Tunnel on April 13, 1996. Fortunately, the train was successfully switched onto a siding designed to stop runaways.

Runaways are still common even today, particularly in mountainous Canada. The Transportation Safety Board of Canada reported 159 run-

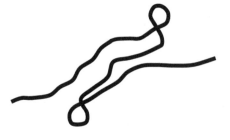

FIGURE 9.2. Canadian Pacific's famous Spiral Tunnels. *Adapted from Gibbons.*

aways during the period 1996–2006. Most were relatively benign events involving a few cars rolling a few hundred feet (km) during switching.

FORCE, TORQUE, WORK, AND POWER

In a nutshell the following is required to operate a train.

Going uphill, the locomotives must supply sufficient force to overcome train resistance and gravity. More force is required for steeper grades. Force is also required to accelerate the train to a practical speed. Going downhill, the brakes must supply enough braking force to cancel gravity.

The torque on the powered wheels must be within a specific range. Too little torque and the train stalls. Too much torque (or too little friction) and the wheels spin. With too little braking torque, the train does not slow in a timely manner. Too much braking torque and the wheels slide, reducing braking effectiveness and potentially damaging the wheels and rail.

Available horsepower determines how fast the train gets up the hill. When climbing a hill, a train limited by horsepower will automatically slow to a speed corresponding to the available horsepower. Going down the hill the train can also be limited by available braking horsepower.

Force

The force required to overcome gravity for a 10,000-ton (9,072-MT) train on a 1% grade has already been described as 1% of the train's weight, or 200,000 lbs (889 kN). An additional 30,000 lbs (133 kN) or so is required to overcome train resistance. Any greater pulling force will accelerate the train uphill per Newton's Law.

An example force for a modern locomotive is the Electro-Motive SD70 with 109,150 lbs (485 kN) of continuous tractive effort. Just as with multiple people pulling on a rope, the force of multiple locomotives connected in series is additive. Too much locomotive force can break a coupling, as illustrated by the following accident.

In 1996, a 15,600-ton (14,152-MT), 114-car coal train broke in two between the 4th and 5th locomotives going up a 1% grade.[5] The automatic emergency brakes successfully stopped the train. Two helper locomotives were sent to help restart the heavy train, now stuck on the hill.

At $-35°F\ (-37°C)$ the brake system's rubber seals leaked pressurized air. The leakage loosened the emergency brake's grip, and the train became a runaway.

When warned of the runaway headed their way, the train crew of the helper locomotives successfully jumped. The runaway shoved the two helper locomotives backward about 1 mile (1.6 km) before derailing 66 cars and the 2 locomotives. There were no injuries. The National Transportation Safety Board (NTSB) determined that the train was operating too close to the broken coupler's rated strength of 400,000 lbs force (1,779 kN). The accident could have also been prevented if the train crew followed operating rules and applied hand brakes when the train stopped.

Braking Force

Braking forces slow the train per $f = m \times a$. Each brake shoe rubbing on a wheel creates a force opposing the train's motion. A realistic number for the braking force on a car with 8 wheels is 8×850 lbs/wheel $= 6,800$ lbs (30 kN). The 6,800-lb braking force acts to slow the car as if lifting a 6,800-lb weight, as shown in Figure 9.3. Since the rolling resistance of the car is insignificant, compared to the braking force, the car acts like a block sliding on a frictionless surface.

Torque

A 20-lb (88.9-N) force on the end of a 10-inch (0.254-m) wrench supplies a torque of 10 inches $\times$ 20 lbs $= 200$ inch lbs (22.59 mN).

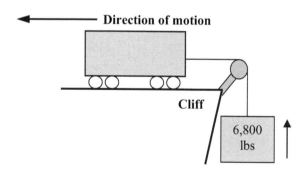

FIGURE 9.3. The 6,800-lb retarding force on the car's 8 brake shoes have the same effect slowing the car as lifting a 6,800-lb weight.

Torque = force × length. One way to loosen a stubborn nut is to create a larger torque by applying the same force using a longer wrench.

The force to lift a train up a hill must be supplied as a torque through the powered axles. If a nut is attached to the locomotive's wheel, the train will move uphill if the nut is turned with enough torque. Insufficient torque and the locomotive's wheels stall. All trains can be made to stall going uphill by adding more cars until the locomotive's wheels refuse to rotate.

Our example locomotive with 109,150 lbs (485 kN) of continuous traction effort has 6 powered axles. The 109,150 lbs of force divided by 12 wheels equals 9,096 lbs (40.5 kN) per wheel. If each wheel has a radius of 21 inches (0.5 m), the torque per wheel equals 9,096 lbs × 21 inches = 191,000 inch lbs (21,580 mN) of torque per wheel.

Powered Wheels Can Slip

It's actually frictional forces that push a train forward. If there is adequate torque and power to pull the train uphill, but not enough friction, the wheels slip.

To see how the frictional force moves the train uphill, replace a powered axle with a rotating arm that pushes off of a bar welded to the rail, as shown in Figure 9.4. The welded bar's force on the rotating arm imitates the frictional force. The locomotive is held in place with a chain attached to a rigid post.

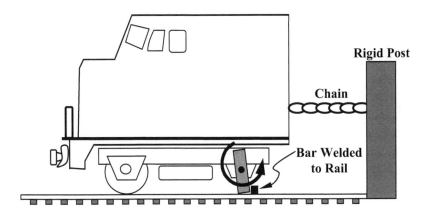

FIGURE 9.4. Replacement of a powered wheel with a torqued arm pressing against a bar welded to the rail.

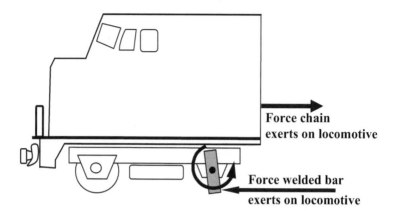

FIGURE 9.5. Horizontal forces on the locomotive in Figure 9.4. Frictional force pushes the locomotive forward.

Figure 9.5 identifies the horizontal forces holding the locomotive in equilibrium. If the force pulling the chain is replaced by a force pulling a train, the frictional forces are seen to be the only forces available to move the train forward.

If the torque on the arm increases, the opposing force from the welded bar automatically increases within limits, as does the pulling force on the chain. There is a maximum torque the bar welded to the rail can withstand before it breaks. Similarly, there is a maximum torque before friction loses its grip. If the torque exceeds the frictional grip, the wheel spins. In Figure 9.4 excess torque on the powered wheel will cause the wheels to spin. Frictional heat from spinning wheels can damage the wheels and/or rail.

The maximum frictional gripping force increases with the weight on the wheel. A wheel with twice as much weight on it will have twice the available frictional grip. Similar to adding weight to the trunk of an automobile in the winter to improve traction, the weight on the locomotive's powered wheels is a fundamental design factor that determines available pulling force. This relationship is described with the coefficient of friction.

maximum frictional force = coefficient of friction
× weight on a powered wheel

Realistic values of the coefficient of friction might vary between 0.15 for frosty rails to 0.30 for clean, dry rails. (The coefficient of friction is also known as the factor of adhesion.)

To climb a 2% grade a 10,000-ton (9,702-MT) train must have at least 430,000 lbs (1,912 kN) of frictional force under the locomotive's wheels to overcome the train's resistance and gravity forces. If there is not enough weight on the locomotive's wheels (or not enough friction between the wheels and the rails), the wheels will spin. If the rails become wet, frosty, or greasy, the wheels can spin just like an automobile's wheels do on ice.

Our example locomotive has 373,710 lbs (169,512 kg) on its powered wheels. If the coefficient of friction reduces to 0.25, the locomotive cannot pull more than 0.25 × 373,710 lbs = 93,427 lbs (415 kN), no matter how powerful the engines are. Since our example locomotive can pull more than 109,150 lbs (485 kN), the locomotive wheels will spin when the torque exceeds a critical value.

The weight on each of the locomotive's 12 powered wheels is 373,710/ 12 = 31,142 lbs (138 kN). The maximum frictional or adhesion force (assuming a coefficient of friction of 0.25) each wheel can react against is 0.25 × 31,142 lbs = 7,785 lbs (34.6 kN). If each powered locomotive wheel has a radius of 21 inches (0.53 m), the maximum torque per wheel is 7,785 lbs × 21 inches = 163,500 inch lbs (18,473 mN). If the locomotive's motors try to supply more than 163,500 inch lbs of torque to each wheel, they will spin.

If the wheels begin to spin, the engineer can back off the torque until spinning stops, or he or she can add sand to the rails to increase the coefficient of friction. Sand might increase the coefficient of friction from 0.25 to 0.30. With a coefficient of friction equal to 0.30, the frictional force at the wheels exceeds the locomotive's capacity to pull and the wheels stop spinning.

The very latest alternating current (AC) motors with computerized slip control obtain higher adhesion, up to 1.5 times more effectively than non-computer-controlled wheels. In that case the typical adhesion factor increases to 1.5 × 0.25 = 0.37. The computer control allows the wheels to operate closer to the limit of slip. If one wheel starts to slip, the computer will automatically increase the torque to the non-slipping wheels.

Of course, the actual adhesive force obtained by any locomotive depends on the condition of the rails.

If the wheels are slipping and suddenly grab, the train will lurch and potentially break a coupling. The engineer will try to back off the throttle before coupling failure, and today microprocessors will lessen the jerk. If the engineer increases the throttle, the increased torque and the frictional force (assuming the wheels don't slip) accelerate the train per f = m × a. The train continues to accelerate until its speed is limited by the available horsepower for that throttle setting.

Braking Torque

A 10-inch (0.254-m) diameter brake cylinder pressurized to 50 psi (344 kPa) creates a force of about 3,900 lbs (17 kN). The brake cylinder rod is connected to all 8 brake shoes on each car via the brake-rigging. The brake-rigging multiples the brake cylinder by 8 with levers. Because of rubbing in the linkages, only about 65% of the input force makes its way to the brake shoes. A realistic force pressing each brake shoe against each wheel is about 8 × 3,900 lbs × 0.65 divided by 8 wheels = 2,500 lbs (11 kN) per wheel.

The coefficient of friction for a brake shoe decreases with increasing speed. Example values are about 0.48 for zero mph and 0.27 at 90 mph (145 km/h). A realistic value for a brake shoe coefficient of friction at 20 mph (32 km/h) is 0.34. The retarding force per brake shoe equals 2,500 lbs × 0.34 = 850 lbs (3.8 kN), as shown in Figure 9.6.

The retarding force from the brake shoe acts on the rim of the wheel. A similar force would act on the rim of the wheel if a rope is wrapped around the wheel and lifts an 850-lb force hanging off a cliff, as shown in Figure 9.7. Since the 850-lb weight is being lifted at the same speed the car is moving to the left, it is the same retarding force as a weight directly attached to the car, as shown in Figure 9.3.

The brake retarding force must be large enough to slow the wheel but not so great that it causes the wheel to stop spinning and slide on the rail. Sliding reduces the frictional grip between the wheel and rail and reduces the wheel's ability to slow the train. A sliding wheel rapidly leads to flat spots on the wheel. The flat spot impacts on the rails with every

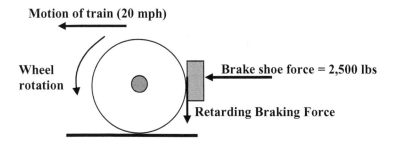

FIGURE 9.6. Brake cylinder force presses on the brake shoe, creating retarding force and torque that slows the train. Force shown is for one brake shoe acting on one wheel.

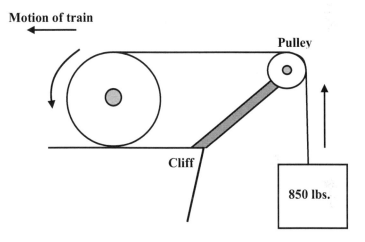

FIGURE 9.7. The braking retarding force acts like a rope wrapped around the wheel lifting a weight. Force shown is for one brake shoe acting on one wheel.

rotation, potentially damaging the rails and/or bearings. (See "Wheel Impact Loads" in Chapter 11.)

Work

Work, an official scientific term, equals force × distance. A person does 200 foot lbs (271 J) of work if he or she lifts his or her 200-lb (90.7-kg) body up 10 feet (3 m). The same amount of work against the force of gravity occurs if the person climbs a ladder straight up 10 feet or walks 20 miles (32 km) along an extremely gradual incline that rises 10 feet.

Obviously, it takes more work to walk 20 miles if one includes the work of swinging your arms and legs. Swinging your limbs burns calories but is not included here. (Technically, the work done by swinging limbs cancels with every back-and-forth motion.) More accurately stated, ignoring friction, a machine (or locomotive) does the same amount of work against gravity lifting a weight 10 feet (3 m), whether the lift is straight up or gradual.

The 10,000-ton (9,072-MT) train (20,000,000 lbs [9×10^6 kg]) lifted straight up 400 feet (122 m) takes 400 feet × 20,000,000 lbs = 8,000,000,000 foot lbs of work (10,800 MJ). It takes additional work to overcome train resistance. The resistance for a 10,000-ton train was previously described as 30,000 lbs (133 kN). If the 10,000-ton train is raised 400 feet by being pulled on a 40,000-foot (12.2-km)-long 1% grade, it will take an additional 40,000 feet × 30,000 lbs = 1.2 billion foot lbs of work (1,626 MJ) to overcome train resistance. The total work to lift the train 400 feet on a 40,000-foot-long 1% grade is 8 billion + 1.2 billion = 9.2 billion foot lbs of work.

Work equals energy. We can also convert 9.2 billion foot lbs of work to 3.46 million watt-hours or the amount of energy from burning 34,640 100-watt bulbs for 1 hour.

Power

Power is the rate of work, or how fast the work occurs. It takes twice as much power for the 200-lb (90.7-kg) person to lift her or his body 10 feet (3 m) in 10 seconds than it does to accomplish the same feat in 20 seconds.

Power equals work per unit time or:

$$\text{Power} = \frac{\text{Force} \times \text{Distance}}{\text{Time}}$$

If the force remains constant, then:

$$\text{Power} = \text{Force} \times \frac{\text{Distance}}{\text{Time}} = \text{Force} \times \text{Speed}$$

In one version of the story, James Watt was trying to market his steam engine in eighteenth-century England. After observing the competition, a horse, he somewhat arbitrarily defined 1 horsepower as follows:

$$1 \text{ horsepower} = 330 \text{ lbs} \times \frac{100 \text{ feet}}{\text{minute}}$$

In other words, Watt estimated that 1 horse could lift 330 lbs (150 kg) up 100 feet (30.5 m) in 1 minute (Figure 9.8). Assuming the same horse mechanically "works" at a constant power output over a variety of speeds, the horse would also supply 1 horsepower (0.74 kw) by lifting 165 lbs (75 kg) 100 feet (30.5 m) in 0.5 minute, or any other combination that works out to 33,000 foot lbs per minute (746 watts).[6]

More power pulls a train with more speed. Theoretically, a locomotive could be designed to pull a 10,000-ton (9,072-MT) train up a 1% grade using only 5 horsepower (3.7 kw)—if 0.008152 mph (0.013 km/h) is an acceptable speed.

The pulling force, or tractive force, of a locomotive determines how much tonnage can be pulled up a given grade. The rated horsepower determines how fast the tonnage goes up the hill. The required horsepower is a very important operating parameter for a railroad and depends

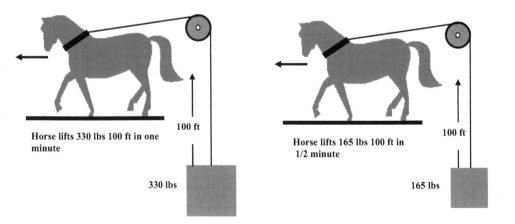

Horse lifts 330 lbs 100 ft in one minute

100 ft

330 lbs

Horse lifts 165 lbs 100 ft in 1/2 minute

100 ft

165 lbs

FIGURE 9.8. Both horses are working at the same rate and produce 1 horsepower.

on how much tonnage it wants to haul, how fast it wants to haul it, and the ruling grade.

Railroads target required horsepower based on their operating conditions. On relatively level ground, a slow 15,000-ton (13,608-MT) coal train might only need 0.5 horsepower (0.37 kw) per ton, while an 80-mph (129-km/h) intermodal freight train may require 2.5 horsepower (1.47 kw) per ton to maintain speed. The biggest locomotives ever built (steam or diesel) were about 6,000 horsepower (4,474 kw). Today locomotives with 3,000 to 4,400 horsepower (2,210 to 2,940 kw) are most common.

The locomotive may have enough tractive effort to pull the train up the hill and enough horsepower to move it fast enough, but the locomotive will stall on the hill if the wheels are slipping. The wheels will slip if more torque is applied to the wheels than can be supported by the frictional force between the wheels and the rails.

Rotated Shaft Transmits Horsepower

A horse pulling at a given force and speed supplies horsepower. So does a shaft rotating with a given torque and rotational speed. In Figure 9.9 the input horsepower is the horse pulling the rope. The output horsepower is the weight being lifted up. The input power equals the output power. The rotating shaft must transmit the same amount of horsepower. If the horse increases its speed, the rotational speed of the shaft also increases. If the horse increases its pulling force, the torque on the shaft also increases. In both cases horsepower increases. Just as power = force × velocity, power also = torque × rotational speed.

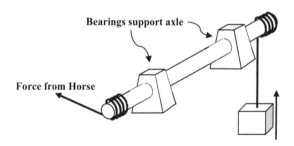

FIGURE 9.9. The power of a horse pulling a rope can also be described in terms of torque and rotational speed of an axle.

Frictional Power

Consider a block pulled on a horizontal surface by a horizontal force at 20 mph (32 km/h). If the horizontal force equals the frictional force, the block continues indefinitely at 20 mph. The frictional force is a fraction of the vertical force, the block's weight. This fractional value is also called the coefficient of friction between the block and surfaces (Figure 9.10).

The coefficient of friction for steel sliding on steel varies greatly, depending on the condition of the surface. If the coefficient of friction is 0.25 and the block weighs 400,000 lbs (181,400 kg), the frictional force equals 0.25 × 400,000 lbs = 100,000 lbs (445 kN).

If the block is dragged with a force of 100,000 lbs at a speed of 1,000 ft/min (11.36 mph [18.28 km/h]) across a steel surface with a coefficient of friction of 0.25, the required power to pull the blocks equals 3,030 horsepower (2,260 kw).

It takes 3,030 horsepower to drag a 400,000-lb (181,400-kg) steel block across a steel surface at 11.36 mph—very close to the rating of our example locomotive used previously. Our example locomotive is actually rated to pull 109,150 lbs (485 kN) continuously at 11.4 mph (18.34 km/h) equal to 3,318 horsepower (2,474 kw). It does this with a diesel engine rated at 4,000 horsepower (2,982 kw). The extra horsepower is used to pull itself and to power the lights, airbrake compressors, blowers, and the like. (The same locomotive could easily pull 66,000,000 lbs [29,900 MGT] on level ground with steel wheels used on steel rails.)

The 400,000-lb block being pulled with 3,030 horsepower on a steel

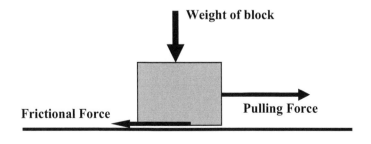

FIGURE 9.10. Forces on a block sliding on a table.

surface also dissipates 3,030 horsepower of frictional power. From the conservation of energy the frictional energy reappears as heat energy that accumulates and slowly raises the block's temperature. Increasing the speed or weight of the block generates more heat. The same process occurs when rubbing your hands together. Press your hands together harder or move them faster and more frictional heat is created.

If the block is pulled by a horizontal force greater than the frictional force, the block accelerates per f = m × a. The block cannot accelerate indefinitely, but will accelerate until limited by the available horsepower.

Braking Power

The brake cylinder presses the brake shoe against the wheel. This creates a retarding force and torque that dissipates braking power.

If the train is moving at 20 mph (32 km/h), the rim of the wheel slides past the brake shoe at 20 mph. A 850-lb (3.8-kN) retarding force on one wheel (see Figure 9.6) moving at 20 mph creates 45 braking horsepower (33 kw) for 1 brake shoe trying to stop the train. All of this frictional power becomes heat energy available to heat up a typical 765-lb (347-kg) wheel and 8-lb (3.6-kg) brake shoe.

When the average wheel temperature exceeds 500°F (260°C), the frictional force of the brake shoe begins to degrade in a process known as brake fade. The temperature the brake shoes reach depends on the wheel speed, the force pressing on the brake shoe, and, most important, how long the brakes are applied. The heat-up effect is cumulative. If the brakes are applied, released, and reapplied (before cooling off), the brakes will continue to heat up. Accidents with brake fade are discussed in the following chapter.

TOO HEAVY, TOO FAST

A runaway can also occur if the train has too much kinetic energy—the energy of motion.

$$kinetic\ energy = 0.5 \times mass \times velocity^2$$

If the train's kinetic energy exceeds the brake shoe's ability to dissipate heat, the brakes fail by overheating and brake fade. The train kinetic energy can be excessive if the train's speed or mass or some combination of the two is too much for the brakes to handle in the conditions present.

The 1917 runaway briefly described in Chapter 2 can now be explained as a heavy train with too much kinetic energy for the braking system.

TRAIN TOO HEAVY: MODERN VERSION

Southern Pacific Train 7551 started out as 4 locomotives in Mohave Yard, California, on May 12, 1989. The goal was to transport 69 hopper cars filled with soda ash (used to make glass, detergents, and fertilizer) to the Port of Los Angeles. Because one of the 4 locomotives would not start, the dispatcher decided Train 7551 would meet and attach 2 additional locomotives—a helper unit. The additional locomotives were needed to climb Cajon Pass (elevation 4,190 feet [1,277 m]) between the San Bernardino and the San Gabriel Mountains.

The train crew moved the four original locomotives repeatedly back and forth and on and off the main track to pick up their train, reposition the locomotives at the head of the train, move off the main track for oncoming traffic, and couple to the helper locomotives. This part of the story is told to appreciate how trains, sometimes nearly 2 miles (1.2 km) long, can experience many movements with potential for human error and to also explain what keeps the dispatcher busy.

After a required initial airbrake test, Train 7551, now just 4 locomotives, departed Mohave Yard at 12:15 a.m. Because the pickup siding for the soda ash, just 3.7 miles (6 km) away, was occupied by track repair equipment, the dispatcher instructed them to proceed 6 miles (10 km) past their pickup point and clear the main track for oncoming traffic. After a 35-minute wait, Train 7551 retraced its route to the pickup point and coupled the 69 loaded cars. Because the siding still had equipment on one end, the 4 locomotives coupled to the back of the train. The train continued back to Mohave Yard to switch the locomotives to the front of the train. More than 3 hours after the original departure, the 4 locomotives were back where they had started. After another airbrake test, the 69 cars and 4 locomotives departed Mohave Yard for the second time at 3:35 a.m.

Just 19 miles (30.6 km) from Mohave Yard, Train 7551 moved onto a siding and waited for the helper unit. After the 2 helper locomotives coupled to the back of the train, yet another mandated airbrake test was performed. The now complete train of 69 cars and 6 locomotives finally moved back onto the main track at 5:30 a.m. (Every time cars are coupled or uncoupled an airbrake test is required to verify continuity of the airbrake line; see "Brake Line Blocked" in Chapter 10.) It had taken Train 7551 more than 5 hours to move just 19 miles (30.6 km).

A modern diesel electric locomotive uses a diesel engine to generate electricity that powers electric motors attached to each axle.[7] If the current is reversed in the electric motors, they convert into generators that extract energy from the wheels and slow the train. Modern trains have airbrakes and electric brakes called dynamic brakes. Both braking systems will be explained in more detail in the following chapter.

Of the original four locomotives, one was not functioning and the dynamic brake on a second unit was noticed to operate intermittently. Also, the dynamic brake on one of the two helper locomotives was not working. Of the six locomotives, only three had fully functional dynamic brakes. The engineer in charge believed there were four. The most serious mistake associated with Train 7551, however, had occurred more than a week earlier. A clerical error listed the weight of Train 7551 as 6,150 tons (5,579 MT) instead of 9,000 tons (8,165 MT).

THE RUNAWAY

The train crested a hill (milepost 463) 83 miles (133 km) from Mohave Yard at 25 mph (40 km/h) at 7:03 a.m. and began a partial airbrake application before descending the 22-mile (35-km)-long, 2.2% grade hill. To apply the brakes, the system air pressure is decreased; exactly how this works is explained in the following chapter. This is done as a safety feature. If the train comes apart for any reason, the pressure reduction automatically applies the emergency brakes, stopping both parts of the train and preventing a runaway.

There is only so much air pressure available for reduction before the airbrakes stop working. The engineer reduces the air pressure in the brake line incrementally until he or she eventually reaches the 26-psi (179-kPa) maximum reduction, known as a full service reduction.

While cresting the hill at 25 mph (40 km/h), the engineer applied a

6-psi (41-kPa) brake pipe reduction. The train accelerated from 25 to 30 mph (40–48 km/h) and the lead engineer increased the brake pipe reduction to 10 psi (69 kPa) and then to 14 psi (96.5 kPa). Six miles (9.6 km) down the hill, the engineer increased the brake pipe reduction to 18 psi (124 kPa); the train was traveling at 31 mph (50 km/h) and still accelerating. The engineer still considered the situation under control, but began to worry after reaching a brake pipe reduction of 20 psi (138 kPa).

Southern Pacific operating rules dictate that a train must stop immediately (with the emergency brakes) if it cannot be controlled with half of a standard full brake pipe reduction. The emergency brake system contains a separate source of pressurized air available to prevent runaways. This train should have been stopped when the brake pipe reduction exceeded 13 psi (89.6 kPa) at milepost 467, just 4 miles (6.4 km) down the 22-mile-long hill.

The lead engineer and the helper engineer, the road foreman, and the general foreman, however, all testified that they considered the "stop the train when brake pipe reduction exceeds 13 psi" as a recommendation and not a mandatory rule. They all agreed that engineers routinely violate this limit.

The lead engineer reached the 26-psi full brake pipe reduction limit at milepost 477. The train was still accelerating. At 7:30 a.m. the train reached 45 mph (72 km/h) and the helper engineer in the back applied the emergency brake. The brake shoes became overheated and started to smoke. A passing motorist reported the train being engulfed in a light blue smoke.

At 7:33 a.m. (milepost 481) the conductor from the front of the train radioed the yard master: "We have a slight problem, I don't know if we can get this train stopped." The helper engineer, in the back of the train, had different thoughts. He braced himself on the floor and radioed at 7:37 a.m., "Mayday! Mayday! . . . we're doing 90 mph (145 km/h), nine zero, out of control, won't be able to stop till we hit Colton (milepost 494)."[8] Seconds later, the train was on the ground at milepost 486.6. The entire train was wrecked after plunging down a 20-foot (6-m)-high embankment[9] and slamming into a row of houses. Seven homes were destroyed and four seriously damaged. It was later estimated that the train's speed exceeded 100 mph (160 km/h).

SOUTHERN PACIFIC OPERATING RULES

The maximum speed limit, set by Southern Pacific, was 65 mph (105 km/h) for freight trains on this track. There are special rules, however, for Cajon Pass.

A heavy train speeding too fast downhill can become a runaway, similar to a person being out of control when running too fast down a hill. If the train contains more energy of motion, or kinetic energy, than the brakes can dissipate as heat, the brakes will overheat and lose their effectiveness.

Because of the velocity-squared term in the kinetic energy equation, a train traveling 60 mph (97 km/h) has 4 times as much kinetic energy as a 30-mph (48-km/h) train and 16 times more than a 15-mph (24-km/h) train. One method to safely descend the hill is to reduce the speed and limit the train's downhill kinetic energy to within the capability of the brakes.

Depending on the weight of the train and the number of operating dynamic brake axles (there are six axles of dynamic brakes on most locomotives), Southern Pacific rules provide for a variety of speed limits on this hill. Above certain limits, no train is permitted to descend this dangerous hill.

The engineer believed he had 69 tons per car trailing weight and 256 tons per axle of dynamic brake.[10] For those conditions the speed limit descending this hill is 30 mph (48 km/h). Considering the correct weight and the inoperative dynamic brakes, the train in fact had 130 tons per car and 500 tons per dynamic brake axle. The operating rules should have stopped this train from descending this hill. The speed limit for various combinations of train weight and dynamic brakes for Cajon Pass are given in Table 9.3.

THE IMMEDIATE AFTERMATH OF THE DERAILMENT

The lead locomotives rolled over onto their left sides with substantial crushing damage. The surviving lead engineer was seated on the right, the fatally injured conductor was seated on the left. The brakeman was also found dead on the left side of the third locomotive.

TABLE 9.3

Operating Rules for Cajon Pass

Weight per Car	Dynamic Brakes per Axle Do Not Exceed:	Speed Limit (mph)
Fewer than 80 tons/car	Dynamic brakes not required	30
Fewer than 125 tons/car	225 tons per axle	30
	300 tons per axle	25
	400 tons per axle	20
Between 125 and 140 tons/car	225 tons per axle	30
	300 tons per axle	20
Greater than 140 tons/car	Greater than 300 tons per axle	Train not permitted to operate

A resident called 911 about 7:41 a.m. The San Bernardino fire battalion chief arrived 7 minutes later. Concerned about the white powdery substance spilled from the train, the chief requested a hazmat unit. The fire chief testified that he had been aware of a gasoline pipeline buried near the tracks but had been uncertain exactly where. The fire department began a house-to-house search for survivors; shortly after, two children were found dead inside a destroyed house.

The deputy fire chief arrived around 8:05 a.m., set up a command post, and assumed control of the emergency situation as the onsite incident commander. The incident commander has legal authority to protect the public's health. Pipeline and rail operations cannot resume until authorized by the incident commander.

A crane was set up to remove the wrecked cars. Meanwhile, a resident was reported missing and nothing was moved until the search dogs completed their work. After removing four truckloads of soda ash, train, and house debris, emergency workers discovered a void in the rubble. Shortly after nine o'clock p.m., a fireman stuck his hand in and felt another hand squeezing back! The dogs had discovered a survivor under a collapsed house, train wreckage, and 12 feet (3.6 m) of soda ash. Said one of the rescuers, "I would never have believed this individual could have survived where he was. He had major train parts against his body."[11]

Running along the length of the track was a 14-inch (0.35-m) gasoline pipeline. Flow was stopped at 8:30 a.m. Normally, the pipeline operates with a pressure of more than 1,600 psi (153,200 kPa). With the pumps shut off, the residual pressure was 1,128 psi (108,000 kPa).

The train derailed 0.1 mile (161 m) short of a check valve that was 6.9 miles (11 km) from the pumps. The 6.9 miles of pipe contain almost 266,000 gallons (1 million l) of gasoline. At 1,128 psi, the gasoline compresses slightly, enough to make room for an additional 17 gallons (64 l) of gasoline inside the 6.9-mile-long, 14-inch pipe. In other words, if the pipeline is completely filled at zero pressure with 266,000 gallons of gasoline, it would take 1,128 psi to squeeze an additional 17 gallons in. The plan was to remove at least 17 gallons of gasoline to reduce the pressure.

Workers removed 5,000 gallons (18,900 l). The pressure in the pipeline did not drop. Four check valves were located at 6.9, 14.9, 19.2, and 25.7 miles (11, 24, 31, and 41 km) from the pumps and up toward the top of Cajon Pass. A check valve swings opens to allow flow in one direction, but then swings closed to prevent backflow in the opposite direction, as shown in Figure 9.11.

The pipeline workers thought the backflow was too slow to cause the flapper to swing shut. They tried pumping out an additional 5,000 gallons at a faster rate. The workers tried a third time to reduce the pressure of the gasoline inside the pipeline and then gave up.

The pumps (elevation 1,040 feet [317 m]) pump the gasoline up 28 miles (45 km) of pipe over the Cajon Pass (elevation 4,480 feet [1,365 m])

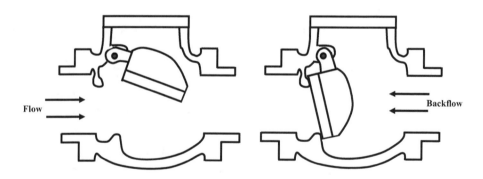

Flow

Backflow

FIGURE 9.11. One of many configurations for a check valve. When the flapper is open, flow proceeds from left to right. When the flapper is closed, flow in opposite direction is prevented.

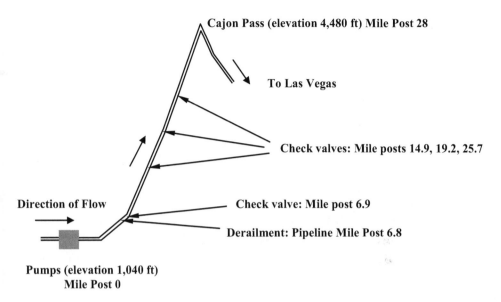

FIGURE 9.12. The check valve at just past the derailment site should prevent 28 miles (45 km) of pipe running up the mountain from draining into any potential leak.

and eventually to Las Vegas, as shown in Figure 9.12. If the pipeline leaks at the bottom of the mountain, 28 miles (45 km) of gasoline will drain from the top of Cajon Pass and squirt out at a pressure of about 1,100 psi (7,584 kPa). (The pressure at the bottom of a column of gasoline 4,480 feet − 1,040 feet = 3,440 feet high [1,048 m] is about 1,100 psi.)

The flapper check valves are there to prevent backflow of the gasoline from the top of Cajon Pass into any potential leak at the bottom, not just any leak, but a high-pressure 1,100 psi leak.

Later, the investigators decided that the first two check valves had been stuck open, an important fact for the remaining story yet to unfold. The pipeline workers should have considered this possibility when they failed to reduce the pressure in the pipeline.

Meanwhile, back at the derailment site, broken train parts from the 69 hopper cars and 6 locomotives were strewn about with enough soda ash to cover a football field 27 inches (0.68 m) deep. Many cranes, bull-dozers, backhoes, excavators, and front-end loaders were very active at the derailment site, cleaning up the mess.

Workers dug several holes to locate the pipeline and several trenches to partially expose the pipe for visual inspection. The pipeline was buried 4–8 feet (1.2–2.4 m) deep. All large chunks of metal penetrating the ground over the pipeline were carefully investigated. Also, Southern Pacific was careful to lift (instead of drag across) any locomotive wreckage on top of the pipeline.

After being assured the pipeline was safe to operate, the deputy fire chief terminated his role as incident commander at ten o'clock p.m. three days after the accident. The excavation/inspection activity occurred feverishly up to the morning of May 16, four days after the derailment. After repairing 600 feet (182 m) of track, railroad traffic resumed at four o'clock p.m. on the 16th. A last-minute threat of an injunction from the city of San Bernardino threatened a delay. The railroad agreed to purchase 11 houses, establish a buffer with the neighborhood, and pay other costs.

The pipeline, supplying more than 2 million gallons (7.5×10^6) of fuel per day to 3 air force bases and 90% of the gasoline used in Las Vegas, was restarted on the 16th at noon with a sense of urgency. The pipeline startup procedure called for slow pressurization and careful observation for any pressure drops indicating leaks.

During the subsequent investigation, the pipeline's maintenance superintendent testified that he had notified the National Response Center, the California Office of Emergency Services, the California State Fire Marshal's Office, and the Underground Service Alert System. All of these agencies and the U.S. Office of Pipeline Safety (OPS) responded to the derailment and participated in the repair and inspection.

THE PIPELINE RUPTURE

On May 25, three 1,000-horsepower (756-kw) pumps were operating at maximum output of nearly 100,000 gallons (378,500 l) per hour and a pressure of 1,620 psi (11,100 kPa). At about eight o'clock a.m. the computer controls shut down the pumps with a low-pressure sensor reading. This did not necessarily indicate a leak, but could have been caused by operating conditions. The pipeline operators tried to restart the pumps for the next few minutes until receiving a call about a fire.

Other workers were instructed to close valves and isolate any pipeline rupture from the rest of the system.

It took workers 55 minutes to drive to pipeline milepost 25.7 and manually shut off the first shutoff valve. The train derailed 0.1 mile (161 m) downstream of the first check valve. If all check valves had worked as designed, only 0.1 mile of gasoline would have drained into the fire. Instead, 25.7 miles (41 km) of gasoline fueled the fire.

Meanwhile, a resident in her backyard reported an unusual white rain falling between 7:45 a.m. and 8:00 a.m. After going back into her house, she heard an explosion before her house caught on fire. Other residents reported seeing a geyser of liquid spouting from where the train had derailed.

The battalion fire chief arrived on the scene around 8:13 a.m. and re-quested that the utility companies shut off electric power and natural gas. He also evacuated 170 people in a 4-block area.

After the pipeline manager arrived, he joined the deputy fire chief in a police helicopter. They saw a stream of burning liquid shooting out of the ground up to 300 feet (91 m). The pipeline manager advised the fire chief to let the fire burn itself out. After destroying 11 houses, the fire did just that around 3:30 p.m. Two residents were fatally burned.

Sparks from a passing train were believed to have ignited the fire. Residents had previously complained about wheels rubbing on the curved track and throwing sparks that ignited small fires. Trainmen would fol-low and extinguish the flames.

The depth of pipe at the rupture site was 2 to 2.5 feet (0.6 to 0.76 m) below ground. Apparently, the cleanup workers, reporting the pipeline depth at 4.5 to 6 feet (1.4 to 1.8 m), had never correctly located the pipeline.

On May 26, the Office of Pipeline Safety ordered the pipeline com-pletely dug up for inspection. On May 30, the OPS determined the pipe-line was unsafe and mandated replacement of 600 feet (183 m) of piping, pressure testing with water to 2,200 psi (15,100 kPa), and repair of all check valves. The new, thicker pipe was encased in concrete. The pipe-line was restarted on June 9, with the mayor of San Bernardino standing on the explosion site as a symbolic gesture of approval.

The pipeline ruptured on top of the pipe with a 29-inch (0.73-m)-long tear. There were also many scratches and dents near the rupture consistent with backhoe damage. The NTSB concluded that the pipeline rupture was from the excavation equipment and not from the derailment.

The fire chief said that he had requested a complete visual inspection of the buried pipe several times, but that the pipeline managers stated they had never received any such request from the fire department or Southern Pacific. Meanwhile, city officials (before the fire) had reached an agreement with the railroad company to pay for digging up the pipeline for complete inspection. Somehow the fire chief was never informed. The onsite incident commander has the authority to demand digging up pipelines; furthermore, he or she can refuse to allow startup of pipelines and railroads if not satisfied with safety standards.

Experts on railroads and pipelines were advising the deputy fire chief (hopefully in good faith) on what was happening and what needed to happen with their equipment. Presumably, the deputy fire chief, having never experienced a similar situation, was a bit overwhelmed.

Risking lives is, of course, unacceptable. When that possibility incorrectly appears remote, however, shutting off gasoline to Las Vegas also becomes a priority.[12] (Gasoline for police, firemen, doctors, and nurses in Las Vegas is important too.) One easy decision—all gasoline pipelines derailed on will be dug up for complete inspection.

The San Bernardino disaster is often studied as an example of one of the worst possible train disasters.[13]

Trains can run away if left unsecured or can run away under power if too heavy or too fast. Trains will also run away because of brake failure, the subject of the following chapter.

10

More Runaways

Brake Failure

The first thing to understand about brake failures is that they are far more common with trucks. In 2006, there were 385,000 large truck accidents, killing nearly 5,000 people. About 30% of the accidents had braking issues.

BRAKE LINE BLOCKED

One of the easiest things to go wrong, and the first thing investigated after a runaway accident, is airline blockage.

A 16-car Pennsylvania Railroad Federal Express passenger train left Boston at eleven o'clock p.m. on January 14, 1953, bound for Washington, D.C. The train stopped uneventfully twice. About 70 miles (112 km) from Boston, the train was delayed with sticky brakes. After inspecting the train for 45 minutes, trainmen found a closed angle cock on the rear end of Car 3.

The airbrakes are activated by a pressurized brake line that runs the length of the train. The brake line must be sealed on both ends of the train. Since either end of any car can be the end car, each car has an angle cock (i.e., shutoff valve) on both ends. Except at the two ends of the train, all angle cocks must be open (Figure 10.1). If any other angle cock is

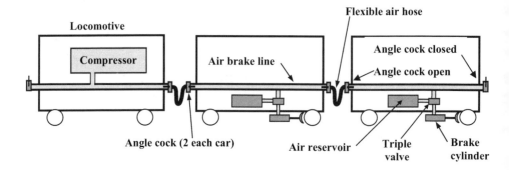

FIGURE 10.1. Components of the automatic airbrake system. Note that the angle cocks must be closed on each end and open elsewhere.

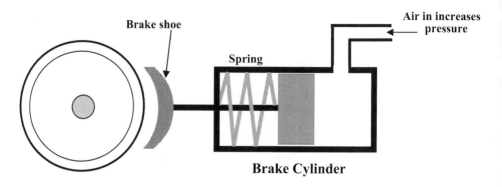

Brake Cylinder

FIGURE 10.2. Increased pressure activates the brake shoe. The spring releases the brake when the pressure is reduced.

closed, air flow is blocked and the brakes are disabled past the closed angle cock.

Airbrakes use air to pressurize a sliding piston inside the brake cylinder. The pressurized piston presses the brake shoe against the steel wheel, as shown in Figure 10.2. Friction between the brake shoes and the wheels slows the train. When the air pressure is reduced, a spring releases the brakes.

Following standard procedure, the Federal Express tested its brakes in Boston and again in New York City. Testing involves applying the brakes and verifying that braking occurs in the back of the train. If braking does not occur, something is wrong—possibly a closed angle cock.

The train switched crews in New York City. The new crew was not informed about the previously closed angle cock anomaly. Routine inspections were made in New York City. The position of the handle clearly indicates if the valve is open or closed. The inspectors insisted that they would have noticed a closed angle cock, something they are trained to look for.

The Federal Express left New York City at 4:38 a.m. The brakes were successfully used to stop at Philadelphia, Wilmington, and Baltimore. The train left Baltimore at 7:50 a.m. and did not use the brakes again for the next 39 miles (63 km), until just 2 miles (3.2 km) from Union Station in downtown Washington, D.C.

Just 1.4 miles (2.2 km) from the station and traveling at an estimated 60–70 mph (97–113 km/h), the engineer noticed that the train was not slowing. He then applied the emergency brakes without response. The engineer tried to reverse the electric motors; the circuit essentially blew a fuse. To make matters worse, the train was descending a 5,500-foot (1,676-m), 0.73% grade.

One might think that the brake line running the length of the train is pressurized to activate the brake cylinders. Not so; in fact, it's the exact opposite. Each car contains a pressurized container of air. In a process explained later in this chapter, a control valve in each car responds to a pressure reduction in the brake line to activate the brakes. Pressure is still required to press the brake shoes against the wheels. The pressure comes from a pressurized air tank in each car, not the brake line. The cleverness of this idea is apparent during an accident. If the train comes apart for any reason, the reduction of air pressure in the brake line automatically activates the brakes in every car.

To activate the brakes, the engineer in the front of the train opens a valve in the brake line that reduces the pressure by letting air out. The pressure in the brake line reduced in the first 3 cars, but was not in the remaining 13 cars. The angle cock behind the third car was closed. The train had three-sixteenths of its normal braking.

There are emergency brake valves on both ends of each car. Opening either valve reduces the air pressure in the brake line and activates the emergency brakes. Two trainmen opened the emergency valve on both ends of Car 3 with no effect. The first three cars already had their brakes on.

Someone had to open an emergency brake valve in Cars 4 through 16. The brakeman in Car 4 was slow to respond. The flagman, in the middle of Car 16, was surrounded by passengers standing in the aisle waiting to get off. Today federal law requires the engineer to either have radio contact with a conductor in the last car or other backup safety equipment.

WITHOUT RADIO CONTACT, the engineer sounded a frantic warning with a series of short whistle blasts, the standard procedure for signaling danger. A quick-thinking trainman in a signal tower 1,700 feet (518 m) from the station heard the distress signal and called the head operator. The head operator frantically called the stationmaster's office: "There's a runaway coming at you on track 16—get the hell out of there!"[1] The stationmaster ran from his office, shouting warnings to 7 workers on the concourse. After crashing through the bumper posts at an estimated 35–40 mph (56–64 km/h), the train destroyed the stationmaster's office and plowed across the concourse. The 475,000-lb (215,456-kg) locomotive fell through the concrete floor and into the basement.

The train separated between the second and third cars; Cars 4, 5, and 6 derailed upright. The locomotive and first two cars were damaged considerably. Typical of many collisions, the passengers in the back of the train were unaware of the impact. The floor was quickly patched. President-elect Dwight Eisenhower arrived in Union Station just three days later for his inauguration.

After the accident, an angle cock was found closed—the same one that had unexpectedly closed outside of Boston. How did the angle cock close? Was it malicious sabotage or mechanical defect? The Senate Committee on Interstate and Foreign Commerce held hearings. Despite widespread evidence to the contrary, many of the witnesses and senators seemed focused on potential sabotage.

Extensive frictional heat damage was found on the wheels and brake shoes of the locomotive and first three cars. Abrasion marks were found on the angle cock handle and on a car frame cross member. The Interstate Commerce Commission (ICC) investigation concluded that normal bouncing and oscillations of the cars had bumped the angle cock handle. The Pennsylvania Railroad added metal guards on the valve handles to prevent reoccurrence.

The vice president of Westinghouse Air Brake Company testified in Senate hearings that the angle cock "could not have vibrated loose."[2] He further said that it had to be deliberately turned and that he had never heard of an angle cock handle turning by itself. This differed from testimony of a railroad foreman, who said he had seen quite a few angle cocks in the closed position.

For a period of time FBI agents and railroad police rode trains between Boston and New York (similar to air marshals on airplanes today). At each stop the agents would jump off the train and guard the angle cocks. Modern equipment and procedures, explained later in this chapter, make sabotage extremely remote today.

THE CONDUCTOR AND THE CABOOSE

Until the mid-1980s or so, every freight train had a conductor riding in a caboose in the back of the train. One of conductor's duties was to monitor brake line pressure. If the engineer in the front of the train reduced the brake line pressure, the conductor verified reduction in the back of the train.

If an angle cock was closed, the air pressure reduction from the head of the train would stop at the closed angle cock. If the conductor in the back of the train opened the brake valve, the brakes would be triggered from the rear of the train to the closed angle cock and the entire train would be safely braked.

On every curve the conductor also watched for overheated bearings and/or stuck brakes that smoked or sparked, a far more common problem. Overheated wheel and bearing detectors installed along the track did away with most of the conductor's duties. After difficult labor negotiations, the caboose and associated crew were eliminated in the 1980s. Lost was the ability to initiate the emergency brakes from the back of the train.

In the 1980s and 1990s, there were numerous runaways caused by a blocked brake line—runaways that could have been prevented by a conductor in the back of the train. Just such an accident happened on February 1, 1996, in California on the famous Cajon Pass.

A 3,200-foot (975-k)-long train weighing more than 5,000 tons (4,535 MT) left Barstow, California, at 1:17 a.m. The train had 4 locomotives and 50 cars. The train was stopped on top of Cajon Pass at 3:40 a.m. for 16 minutes by a signal.

While descending the hill, the train was speeding up—always a bad sign on a steep hill. The trainmen knew they were in trouble when the train reached 18 mph (29 km/h), just 3 mph (5 km/h) over their limit. At about 20 mph (32 km/h), the crew applied the emergency brakes with little effect.

Just 4.5 miles (7.2 km) down the hill, the train overturned and derailed on a curve in excess of 70 mph (113/km/h). The engineer stayed with the train and lived; the conductor and brakeman left the cab, jumped (or were thrown), and died. The 4 locomotives and 45 of the 50 cars derailed. Five of the 14 tank cars were carrying flammable material that burst into flames and burned for 3 days.

Computer simulation predicted that the train required working brakes on just 16 cars to safely negotiate the curve. The investigators concluded that a blockage had occurred between the fifth and ninth cars. Fire damage destroyed all conclusive evidence. The leading theories were an angle cock closed by vandalism, a kinked airline, or some other blockage.[3] The 16-minute stop made vandalism plausible.

Air hoses have been known to kink on a worn spot with excess movement. The draft gear on normal freight cars only allows 1 foot (0.3 m) of movement between cars and is not expected to pinch an undamaged air hose. The fifth car had hydraulic cushioning. Two cars coupled with hydraulic cylinders can move up to 3 feet (0.9 m)—enough to make pinching easier.

Historically, there have been accidents when the train crew forgot to open the angle cock after adding additional cars. With a train of 10 cars, the last angle cock on the 10th car must be closed. If an 11th car is coupled to the first 10, the angle cock on the 10th car must be opened. Federal law mandates a brake test to verify brake line continuity every time cars are coupled or uncoupled. The conductor and the engineer would have had to forget to open the angle cock and to perform the brake test. Operating rules also require a running brake test when approaching dangerous hills. The locomotive's event recorder verified both brake tests had taken place.

Today blockage would be defeated by a two-way end-of-train (EOT) device. A radio-controlled brake valve activates the emergency brakes in the back of the train just as the conductor in the caboose used to. Trains without two-way EOT devices are rapidly disappearing; the federal rules still permit a few exceptions for slow trains or trains running on flat terrain. All freight locomotives ordered since 2001 must have a two-way device. Two-way end-of-train devices were still new in 1996. The runaway train did in fact have a two-way end-of-train device. It was not working, however, and the crew was not trained for unexpected failure.

DEVELOPMENT OF AIRBRAKES

The science of airbrakes is best explained by tracing historical developments.

The very earliest trains used stagecoach technology—foot-activated blocks of hardwood pressed against the wheels by levers. Later, as trains got longer and heavier, locomotive steam pressure pressed the brake shoes against the wheels. After a few cars, the pressure reduced as the steam condensed into water. Hundreds of patents were issued for spring-activated brakes, brakes powered by levers, brakes operated by winding a chain onto a shaft, and many other forgettable ideas.

Hand brakes, still used today to hold a train stopped on a hill, remained the standard for many decades. A hand wheel moves a set of linkages and levers to press the brake shoes against the wheels. After the engineer blew the whistle, brakemen would scamper across the roofs of freight trains with a pick handle to rotate hand brake wheels on multiple cars until the train stopped. Needless to say, running on top of a swaying freight car under the best of conditions, let alone on a windy, icy night, was dangerous. The Pennsylvania Railroad continued limited hand brake use until as late as 1927.

GEORGE WESTINGHOUSE AND THE STRAIGHT AIRBRAKE

In 1868, 22-year-old George Westinghouse (an inventor who successfully competed with Thomas Edison) tested what became known as the "straight" airbrake. Pressurized air, from a compressor in the locomotives, was applied directly to the brake cylinders on each car. The pressurized air cylinder pressed the brake shoes against the wheels, as shown in Figure 10.2.[4]

PRESSURE IS A FORCE PER UNIT AREA (pressure = force/area). A 160-lb (72.5-kg) person standing on one foot with a 2-inch by 2-inch (4 in² [25.8 cm²]) heel exerts a pressure of 160 lbs/4 in² = 40 lbs per square inch (276 kPa). If the same person stands on a smaller heel, 0.5 inch by 0.5 inch = 0.25 in² (1.6 cm²), the pressure increases to 160 lbs/0.25 in² = 640 lbs per square inch, or 640 psi (290 kPa).

Similarly, a pressure exerted on an area results in a force: force = pressure × area. A pressure of 50 psi (345 kPa) on a piston with a 10-inch (25.4-cm) diameter exerts a force of

$$\frac{50 \text{ lb}}{\text{inch}^2} \times \pi(10/2)^2 \text{ in}^2 = 3{,}927 \text{ lbs (17.5 kN) pressing on the brake rigging.}$$

The brake cylinder is attached to the brake rigging with linkages and levers. The effect of the brake rigging, lever action, and rigging efficiency is explained in Chapter 9. The actual braking force pressing each brake shoe against each wheel is about 2,500 lbs (11.1 kN). The total braking force for a freight car with 8 wheels is 8 × 2,500 = 10,000 lbs (44.5 kN).

Stopping distances improved from about 1,600 feet (487 m) with hand brakes at 30 mph (48 km/h) to about 500 feet (152 m) with air-brakes. But trains were limited to only 5 or 6 cars; beyond that it took too long to move the air from the front of the train to the back.

The market success of this simple device in passenger trains was phenomenal. By 1876, nearly 40% of all passenger locomotives (more than 15,500) had the Westinghouse brakes. This rapid market penetration was driven by the public's thirst for passenger safety and Westinghouse's astute business sense. Westinghouse's straight airbrake did not fail safe, however. If the train separated for any reason, the brake pipe depressurized and all braking was lost.

THE WESTINGHOUSE AUTOMATIC BRAKE

Recall that the straight airbrake operates by adding pressurized air in the front of the train to pressurize the airbrake pipe and the

brake cylinders. Westinghouse wanted a design that automatically applied the brakes if the train came apart and released (or reduced) the
pressure in the brake line. Yet he still needed increased pressure in the
brake cylinder to press the brake shoe against the wheel.

In one of those thinking outside the box moments of genius, Westinghouse designed a system that pressurizes the brake cylinder when the
airbrake pipe is depressurized. In other words, reduced pressure in the
airbrake pipe causes increased pressure in the brake cylinder. This required two new ideas. First, air pressure is stored locally in a reservoir
on each car. Second, air flows are managed by a triple valve that responds
to pressure differences between the brake pipe, air reservoir, and brake
cylinder to apply and release the brakes.

If an air hose breaks for any reason, the triple valve on each car activates the brakes by allowing air flow from the air reservoir into the brake
cylinder. If the train comes apart, the individual segments will stop and
prevent an accident like the one described at the beginning of Chapter 9.

THE TRIPLE VALVE

The triple valve is named for the three functions it performs:
(1) it allows air flow between the brake pipe and reservoirs to pressurize
the system, (2) it uses air flow from the reservoir to the brake cylinders
to apply the brakes, (3) and it vents air from the brake cylinder to the
atmosphere to release the brakes. To accomplish these three functions,
the triple valve has a piston that moves in response to pressure differences between the brake line, air reservoir, and brake cylinder. During
movement (and spring-loaded return), the piston covers and uncovers
air flow ports to accomplish its three tasks.

APPLYING THE BRAKES

Initially, there is zero pressure in the brake system. Charging
the system involves pressurizing the brake pipe and the air reservoirs in
each car to 90 psi (620 kPa), the most common freight train air pressure
in current use. It may take 15 to 45 minutes to pressurize the system,
depending on how long and how leaky the train is. During charging the
brake cylinder remains at zero pressure.

After charging, each car's triple valve (Figure 10.3) is in its neutral

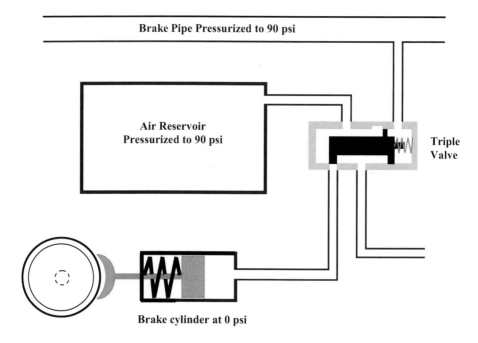

Brake Pipe Pressurized to 90 psi

Air Reservoir
Pressurized to 90 psi

Triple
Valve

Brake cylinder at 0 psi

FIGURE 10.3. After the system is charged, the triple valve is in its neutral position and watching for a brake pipe pressure reduction.

position. The triple valve piston isolates or seals the brake pipe, air reservoir, and brake cylinder pressures at 90, 90, and zero psi respectively. Because the pressure on both sides of the piston equals 90 psi, the piston spring is unstretched.

If the engineer reduces the pressure in the brake pipe from 90 psi (620 kPa) to 80 (551 kPa) (referred to as a 10-lb break pipe reduction), the pressure in the air reservoir now exceeds the pressure in the brake pipe. The triple valve piston responds to the pressure difference and moves to the right, compressing the spring, as shown in Figure 10.4. The movement of the triple valve piston uncovers the port to the brake cylinder. The brake cylinder, pressurized by the air reservoir, presses the brake shoe against the wheel. The pressure in the reservoir decreases as air molecules flow from the reservoir (higher pressure) into the brake cylinder (lower pressure).

Air flows out of the reservoir until its pressure reduces to 80 psi. With 80 psi in the reservoir and brake pipe, the pressure forces are balanced on both sides of the triple valve piston. The spring pushes the triple valve

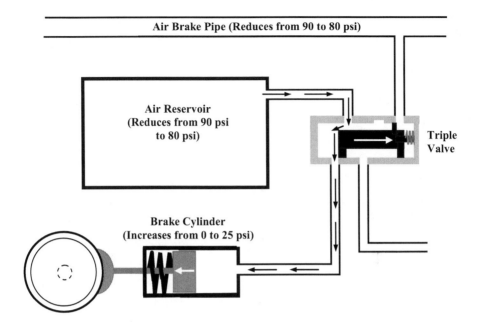

FIGURE 10.4. Apply. The triple valve senses a drop in brake pipe pressure and the piston moves to the right, as shown. Air flows from the reservoir and pressurizes the brake cylinder.

piston back to its neutral position. Because the volume of the reservoir is roughly 2.5 times larger than the brake cylinder, a 10-psi (69-kPa) reduction of the reservoir reappears as a 25-psi (172-kPa) pressure increase in the cylinder. The triple valve at its neutral position seals the pressure in the reservoir and the brake cylinder at 80 and 25 psi respectively. The brake shoes remain pressed against the wheels until released.

Normal brake pipe leakage slowly applied the brakes until the train stopped. To counteract this effect, the engineer had to constantly add air pressure to the brake pipe. An experienced "touch" was required to successfully operate the train and prevent stopping. Since World War II, airbrake systems have automatically adjusted for leakage.

MOLECULES, AIR PRESSURE, AND THE TRIPLE VALVE

Why does the pressure in the cylinder rise 25 psi (172 kPa) when the pressure in the reservoir drops only 10 psi (69 kPa)?

All gas molecules are in constant random motion. This motion results from the molecules' thermal energy and only ceases at absolute zero ($-460°F$ [$-273°C$]). This constant motion explains why a volume of gas immediately fills any container it's placed in. If the gas molecules were not suspended with their thermal energy, they would collect at the bottom of the container like settled dust.

It is helpful to think of a gas as individual molecules. Fundamentally, pressure comes from countless individual air molecules impacting on a surface. The effect of all the impacts averages out and the net effect is uniform pressure. (It is like lying under a plywood sheet constantly pummeled with 5,000 random baseball bounces every second.)

Consider a typical air reservoir with 2,500 cubic inches of volume (41 l). One cubic inch (16 cm³) of air pressurized to normal atmospheric pressure contains about 435×10^{18} air molecules. Add 3×10^{21} more molecules, and the pressure increases to 90 psi (620 kPa). Add twice as many molecules, and the number of impacts doubles (and so does the pressure, to 180 psi [1,241 kPa]). Halve the number of molecules added, and the pressure increases to only 45 psi (310 kPa).

For simplicity, let's say 1 million air molecules removed from the 2,500-cubic-inch reservoir reduce the reservoir's pressure from 90 to 80 psi (from 620 to 551 kPa). The 1 million air molecules added to the brake cylinder increase the number of molecular impacts (and pressure) in the brake cylinder. Because the volume of the brake cylinder is about 2.5 times less than that of the reservoir, the number of impacts per unit volume (i.e., the pressure increase) is 2.5 times higher in the brake cylinder. Hence, a 10-psi pressure reduction in the reservoir will result in a 2.5×10 psi $= 25$ psi increase in the brake cylinder.

As we remove more molecules from the reservoir and place them in the brake cylinder, pressure continues to rise in the brake cylinder by a factor of 2.5 times the drop in the reservoir pressure. When the pressure in the reservoir drops by 26 psi (179 kPa) (from 90 to 64 psi [from 620 to 441 kPa]), the pressure in the brake cylinder will have increased to 26 $\times 2.5 = 65$ psi. The pressure in the reservoir and the pressure in the brake cylinder are now approximately equal. Since air stops flowing between the reservoir and cylinder when the pressures are equal, any further decrease in the brake pipe pressure has no additional effect on the brake cylinder pressure. The maximum braking effort has been obtained.

A brake pipe reduction of 26 psi (179 kPa) is called a full service brake application. After a full service brake application, the engineer has run out of air pressure to brake the train! The engineer can apply the brakes gradually or repeatedly until a total brake pipe reduction up to 26 psi has occurred. At that point no further braking force is available until the brake pipe (and all the reservoirs in all the cars) is replenished with enough air molecules to raise the pressure back up to 90 psi. Unfortunately, if the 26-psi limit is reached on a steep hill, it may be many minutes until the system can be recharged—too late to prevent a runaway.

Full service braking is relative to how much air pressure is in the brake pipe. Starting with less than 90-psi brake pipe pressure will result in a lower full service braking force. Because of leakage along the length of the train, it is also relative to location. Trains are allowed to operate with a brake pipe pressure of 90 psi (620 kPa) in the front of the train and a brake pipe pressure of only 75 psi (517 kPa) in the back.

BRAKE RELEASE

To release the brakes, the pressure is increased in the brake pipe. In this case the triple valve piston senses the increased pressure, moves to the left, and vents the brake cylinder to the atmosphere (Figure 10.5). The brake cylinder pressure decreases to zero. The spring in the brake cylinder returns the brake shoe to its unloaded position. Simultaneously, the flow path between the brake pipe and reservoir is opened. The reservoir is recharged to the brake pipe pressure. In a freight train brake release is all or nothing. There is no gradual release.[5]

The ports to release the brakes are either fully open or fully closed. Because they cannot be partially open, there is no gradual brake release. If a train is going downhill and runs out of pressurized air, the system can only be recharged if the brakes are released—the last thing the engineer wants to do while going downhill!

The engineer can apply the brakes gradually or repeatedly until a total brake pipe reduction up to "full service." The engineer must plan ahead so he or she doesn't run out of air and end up with a runaway train.

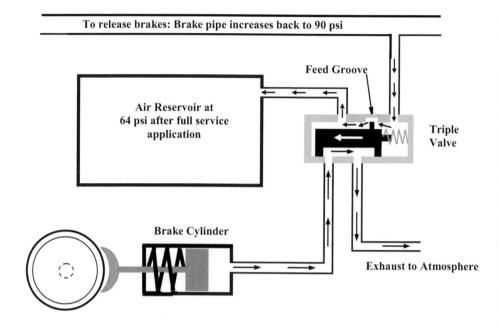

FIGURE 10.5. Release. The triple valve piston senses increased brake pipe pressure and moves to the left. The air reservoir is repressurized, and the brake cylinder is depressurized.

Historically, this has happened many times—so much so that in 1933 a second air reservoir reserved for emergencies was added.

The emergency brakes supply one last blast of air to stop the train. If the train is going too fast and has too much kinetic energy, however, the emergency air supply will not stop the train. Steep grades have reduced speed limits for this very reason.

ADOPTION OF WESTINGHOUSE'S AUTOMATIC BRAKE

The Westinghouse automatic brake was quickly adopted in passenger cars as a matter of public safety. It was also economically easier to place the new technology in the smaller number of passenger cars (at that time there were 25,000 passenger cars compared to 1 million freight cars). Passenger trains, made up of a few cars operated by one company between two fixed points, were easier to update. Freight cars, on the other hand, were shared by many companies and wandered across the nation.

Once tracks and cars became standardized after the Civil War, most routes became interconnected. It became convenient for companies to share cars. The alternative was to unload and reload freight onto a new train every time track ownership changed.

Cars accumulated in yards until there were enough to attach to the next train traveling in the right direction. A car might wander the rail system for years before it returned to its owners. Railroads had lesser incentive to improve freight cars used by another company.

A transition period of years, during which some freight cars had airbrakes and others did not, caused logistic problems. To make the brake pipe continuous, more switching (and switching accidents) was required to group all similar cars together.

Westinghouse, the hardnosed (and extremely successful) businessman, wanted to keep control of his invention. The railroads, familiar with the concept of monopolies, didn't like to deal with one themselves and wanted to buy the automatic brake outright.

The Westinghouse brake was widely adopted on western railroads. There was less freight car sharing in the sparsely populated West, and mountain grades had a greater need for improved braking.

Passing cars between east and west, the Chicago, Burlington and Quincy Railroad was literally in the middle of the airbrake issue. It was annoying to sort out western cars with airbrakes from eastern cars without airbrakes. And all railroads wanted to operate freight trains much longer than 30 cars.

THE BURLINGTON BRAKE TRIALS

All brake inventors were invited to compete and stop a 50-car train during the famous Burlington Brake Trials in 1886. The tests also coincided with Westinghouse's expiring patents. He eventually outsmarted everyone with new patents.

During the first Burlington competition, air pressure reduction in the brake pipe took 18 seconds to reach the 50th car. The rear cars slammed into the stopped cars up front with tremendous shock loads.

A year later, the companies competed again. Every company, including Westinghouse, entered at least one design with instantaneous electric controls. The brake cylinder was still powered by air, but it was activated with electromagnets. Electricity traveled the length of the train at the speed of light.

The electrically activated air cylinders did solve the shock problem—when they worked. Everyone left the trials convinced that electricity had won the day—everyone, that is, except Westinghouse. Westinghouse, who knew a thing or two about electricity, was adamant that electricity would not be reliable for this application.[6]

Electric brake systems had many small and fragile parts subject to damage from shock loads and dirt. During one test, a loose binding caused near total failure. Worse still, it was difficult to find the faulty screw among the 300 connections. (Typically, three wires ran the length of the train with three connections on the ends of each car.)

Problems with electric connectors are well known and include arcing, pitting, loosening, and tarnishing. Gold-plated contacts are used in automotive air bags and many other critical electronic applications. Inert gold remains shiny and does not tarnish.

In 1887, everyone assumed the electric reliability problems would somehow be solved—the shock loading from airbrakes was simply unacceptable. The electric brake dilemma remained unsolved for more than 100 years in freight trains. Shorter passenger trains adopted electric brakes in the 1930s but always with emergency airbrakes as backup.

As the brakes got better, freight trains got longer and re-created the same braking problems. In a classic example of history repeating itself, electronically controlled pneumatic (ECP) brakes (described later in this chapter) are finally becoming established in the twenty-first century. The conversion to ECP brakes remains difficult for all the same reasons.

THE QUICK-RELEASE TRIPLE VALVE

Shortly after the second Burlington Brake Trials in 1887, Westinghouse improved his design and moved the air signal along a 50-car train in just 2 seconds. This ended all competition from unreliable electricity. Westinghouse won the competition and filed new patents.

In a relatively simple modification, the quick-acting triple valve vented air locally out of each valve, instead of only out of the control valve in the locomotive. This greatly increased the propagation of the air pressure signal throughout the train. Hand brakes stopped a 50-car train at 40 mph

(64 km/h) train in 2,500 to 3,000 feet (762 to 914 m). The Westinghouse quick-release airbrake stopped the same train in 581 feet (177 m).

Congress passed the first train safety rules in 1893 requiring a phase-in of airbrakes on freight cars. By 1905, more than 2 million freight cars and 89,000 locomotives in the United States were equipped with the Westinghouse automatic brake. On his death in 1914, a *New York Times* editorial suggested that George Westinghouse's railroad brake had saved more lives than lost in all wars combined.

ENHANCEMENTS TO THE TRIPLE VALVE

The economic incentive of hauling longer trains created a constant motivation to improve the triple valve. Longer trains always posed the same old problem—the possibility that the back of the train will slam into the front during braking. Trains have slowly grown, in great part because of better braking, from 2,000 feet (609 m) and 2,000 tons (1,814 MT) in 1890 to about 10,000 feet (3,048) and 20,000 tons (18,140 MT) today.

There are only two ways to improve the triple valve: speed up brake application and/or speed up brake release. Both are done by manipulating air flow within the valve and between the reservoir, brake pipe, and cylinder.

A 1945 Westinghouse pamphlet shows springs, pistons, ports, passages, slide valves, and chokes added to improve airflow and accomplish enhanced features such as "three-stage quick application," "uniform restricted release," and "emergency quick action." More recently, the pistons that wear have been replaced with rubber diaphragms to improve reliability and simplify repairs.

Improvements continue. (The modern triple valve today is more commonly called a control valve.) The net result is a very fast-acting and sensitive valve—perhaps too sensitive. Response can be erratic from increased leakage in long trains during cold weather.

Today the speed of brake application is getting closer to the limit—the speed of sound in air. An emergency brake application propagates along the length of the train at around 900 ft/sec compared to 500 ft/sec for ordinary braking. This compares to the speed of sound, which varies with temperature and equals 1,011 ft/sec (689 mph [1,109 km/h]) at $-35°F$

FIGURE 10.6. Components of the brake system. These components are often underneath the car and not readily visible.

($-1.7°C$) and 1,177 ft/sec (803 mph [1,292 km/h]) at 110°F (43°C). Components of the brake system are shown in Figure 10.6.

DYNAMIC BRAKES

Except for incremental improvements in the triple valve, braking equipment remained more or less the same until dieselization in the

1950s. When diesel electric locomotives replaced steam locomotives, it became possible to use dynamic braking. Recall that the diesel engine powers a generator that supplies electricity to a motor attached to each powered axle.

The electric motors can also work in reverse as generators that resist the train's motion. This feature is known as dynamic braking, or DB. Switching on the dynamic brakes reverses the electricity flow through the motors. Instead of electricity flowing into the motors, the motors now act like generators and electricity flows out. Advancing the dynamic brake handle produces more electrical energy to slow the train.

Dynamic brakes convert the train's kinetic energy into electrical energy. The electrical energy is in turn converted into heat energy dissipated in a bank of resistors acting like electric heaters. The resistors are very hot and must be cooled by large fans (visible on the roof of most locomotives). Dynamic braking is similar to what is called regenerative braking in hybrid cars except the energy recovered from braking in the latter is used to recharge the batteries.

Because airbrakes cannot be partially released, a train can stall going downhill by overbraking. To avoid stalling, sometimes the engineer wastes fuel by applying the throttle and brakes at the same time. To save fuel DB is the preferred method of braking. DB also saves wear and tear on the brakes and wheels. Excess braking force at the front end of a train, however, can derail it (see Chapter 7).

Depending on the type, dynamic brakes create up to 16,000 lbs (71 kN) per axle of braking force. Because a motor may be not working (or has been purposely shut off to limit braking force in the front of the train), railroads traditionally count the number of effective DB axles.[7]

Figure 10.7 illustrates the total dynamic braking force for a six-axle locomotive, the Electro Motive Diesel (EMD) SD60 with high-capacity, extended-range dynamic brakes. The SD60 diesel engine has 16 cylinders with 710 cubic inches (11.6 l) per cylinder. The SD60 has a maximum tractive effort of 96,300 lbs (428 kN) pulling force at 9.8 mph (15.8 km/h); about 2,500 horsepower (1,864 kw). In comparison, an 81,000-lb (360-kN) dynamic brake force acting at 25 mph (40 km/h) converts to 5,400 horsepower (4,027 kw).

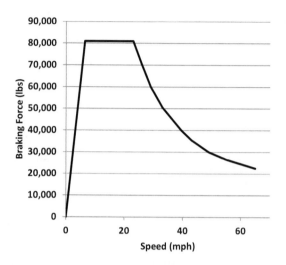

FIGURE 10.7. Dynamic brake force versus speed for an EMD SD60. *Adapted from TSB Report R01W0007.*

RUNNING OUT OF AIR

Since the emergency air reservoir was mandated in 1933, running out of air has been rare. Also, widespread use of dynamic brakes on diesel locomotives has made train control easier going downhill. Accidents, however, still happen.

A freight train with 84 loaded grain cars departed (milepost 0.0) Calgary, Alberta, on December 2, 1997, at 6:52 a.m. The 136-mile (219-km) trip was uneventful for the first 123 miles (198 km). The 5,120-foot (1,516-m)-long train with 2 locomotives weighed 11,350 tons (10,296 MT).

At 11:50 a.m. the train arrived at the top of Field Hill (milepost 123), the start of a 13.5-mile (21.7-km)-long, 2.2% grade. The speed limit downhill was 15 mph (24 km/h). About 2 miles (3.2 km) down the hill, there is a brief stretch of level track, even slightly uphill. At this point, the train was moving at 14 mph (22 km/h) in low throttle with a brake pipe reduction of 7 psi (48 kPa). After resuming the 2.2% descent, the train rapidly accelerated over the next mile in spite of quickly ramping up the dynamic brakes. At maximum dynamic brakes, the train was still accelerating at 25 mph (40 km/h). The engineer made a full service brake application.

When the train slowed faster than expected, the engineer slowly decreased dynamic braking and slowly advanced the throttle. Because of overbraking, the train stalled at milepost 128.1, just 5 miles (8 km) down this difficult hill.

An oncoming train, scheduled to meet the stopped train head on, was to get off the main track and onto a siding at milepost 128.7. Because the stopped train could not see the signal (or the turnoff onto the siding) that coordinated both trains' movements, the engineer planned to roll his train 4,000 feet (1,219 m) by releasing the airbrakes and controlling the train with dynamic braking. After the engineer released the brakes with full dynamic brakes still applied, the train quickly accelerated. At 5.6 mph (9 km/h) the engineer applied a series of small brake pipe reductions without effect.

The brake pipe pressure is lower in the back of the train because of leakage. If there hasn't been enough time to recharge the brake pipe before reapplying the brakes, the next brake pipe reduction must be at least 5 to 8 lbs (34.5 to 55.1 kPa) less than the air pressure in the back of the train to activate all the brakes. The minor brake pipe reductions had no effect other than to use up more air.

The train continued to accelerate until the engineer applied the emergency brakes at 19 mph (30 km/h) to stop the train. The train safely stopped. After a full service brake pipe reduction, quickly followed by a dithering attempt to slow the train and then an emergency braking, the train was out of air.

After the emergency stop, the crew discussed their options. One option was to apply hand brakes on at least half the cars before recharging the airbrake system. Applying the hand brakes, recharging the system, and removing the hand brakes could take a couple of hours.

Another option was to use retainers. Retainers are a choke valve on the brake cylinder exhaust. Recall that the brakes can be gradually applied but only completely released. To recharge the system the brakes must be completely released. Retainers "retain" pressurized air in the brake cylinder, allowing the brake shoes to remain pressed against the wheels while the reservoirs are being recharged. This is the only exception to the statement "the brakes must be completely released to recharge the system." The retainers must be set by hand on each individual car. Retainers, another George Westinghouse invention, are rarely used today except on very steep hills.

The train crew ultimately decided to continue their descent by controlling the train only with the DB. They had used this procedure on previous trips and believed the upcoming curves would also help slow the train. Perhaps their decision was influenced by the prospect of walking the length of the train repeatedly on a cold Canadian December morning.

Applying the emergency brakes automatically set the engines to idle, which also shut off the dynamic brakes. When restarting, the crew used a procedure that worked on other locomotives but not on this newer computer-controlled locomotive. The crew essentially did not correctly reset the computer and started up without dynamic brakes.

After the train brakes were released, the train accelerated to 16 mph (26 km/h) in about 500 feet (152 m). The engineer began making incremental brake pipe pressure reductions totaling 26 psi (179 kPa), but the train continued to accelerate.

The train entered the Upper Spiral Tunnel (milepost 129.0) at about 25 mph (40 km/h) and continued to accelerate to approximately 46 mph (74 km/h) when the emergency brakes were applied; by this time it was far too late. The last 29 cars separated from the 84-car train, derailing 16; the rest of the train continued to accelerate.

The train continued uncontrolled over the next 5 miles (8 km), reaching speeds up to 50 mph (80 km/h). At about 47 mph (76 km/h) on a sharp curve, the 2 locomotives separated and 50 of the remaining 55 cars derailed. The engineer finally recovered the dynamic brakes and safely drove the 2 locomotives to their final destination.

The investigators faulted the railroad company for not having explicit instructions about not going down this hill with a depleted air supply.

BRAKE FADE

Except for unattached cars or running out of air, all other runaways are caused by overheated brakes. The investigators may conclude that the train was too heavy, that it was going too fast, or that equipment failed, but in all cases the brakes overheated.

For a train to slow down, its kinetic energy of motion must be converted into frictional heat energy in the brake shoes. If the heat generated by friction equals the heat dissipated into the air, thermal equilibrium is reached. In that case the wheel and brake shoe heat up to some temperature and remain at that temperature. If the frictional heat generated exceeds the heat dissipated, the 765-lb (347-kg) wheel and 8.5-lb (3.8kg) brake shoe continue to heat up.[8] When the average wheel temperature exceeds 500°F (260°C), brake shoes lose their ability to properly grip, a process commonly called heat fade.

All materials weaken with excess heat. To compensate, the brake shoe can be pressed with a greater force against the wheel. This will increase the braking force, but at the expense of increased heat generation and increased brake fade.

Eventually, the brake shoe and the wheel will show evidence of thermal damage. In extreme cases the steel melts. The molten metal lubricates the rubbing surfaces and significantly degrades braking.

If the brakes start to fade when the train is accelerating down a grade, pressing the shoes harder against the wheel might bring the train under control. At this point, however, there is a race between the increased brake shoe forces trying to control the train and increased heat generation causing more heat fade.

Frictional heat generated depends on the braking horsepower, which in turn depends on the brake shoe force and wheel speed (see Chapter 9). How much and how fast do the brake shoes and wheels heat up? It depends on how hard and how long the brakes are applied and on how much time the brakes can cool off before being applied again. Unless the brakes have time to cool, frictional heat is additive and accumulates.

In the following accident, the investigators focused on heat fade. On the night of January 29, 2000, a yard crew coupled three locomotives. Coupling the locomotives together also involved connecting the air hoses and the multiple-unit cable. The cable allowed all three locomotives to be controlled by one crew in the head unit. Unknown to the crew, a defective cable socket disabled the dynamic brakes in the second and third locomotives.

Immediately after the locomotives were coupled, the airbrakes were tested. Eighty coal cars were connected, and the airbrakes were tested again. The 11,170-ton (10,133-MT) train was ready to leave the yard.

About 2 hours later, the train crew arrived to pick up their train. They performed yet another airbrake test and departed from milepost 281 at 2:30 a.m. The train crew was unable to check the dynamic brakes in the other two locomotives.

At 6:22 a.m. the train started descending a 17-mile (27-km) hill in the Allegheny Mountains of western Maryland with an average grade of 2.4%. The speed limit on this hill is 25 mph (40 km/h).

The train was still in throttle notch 7 (the second highest) from cresting the hill. Over the next 3 minutes, as more of the train draped over the hill, the engineer transitioned from pulling to near maximum dynamic braking. The brake pipe was also reduced 10 psi (69 kPa).

Over the next 7 minutes (from 6:25 to 6:32 a.m.), the engineer made a series of small brake pipe reductions totaling 17 psi (117 kPa). To avoid overbraking and stalling the train, the engineer increased the throttle up to notch 6 for about 2 miles (3.2 km) while traveling downhill at 21 mph (34 km/h).

Sixteen minutes into the descent (just 6 miles down the hill), the engineer switched from power braking to heavy use of the partially working dynamic brakes. Over the next 2 minutes, the engineer maxed out the brakes with a full service reduction of 26 lbs (179 kPa). At 34 mph (55 km/h) and 9 miles (14.5 km) down the 17-mile (27-km) hill, he applied the emergency brakes. It was too late; the brakes were overheated. Applying the emergency brakes also shut off the working dynamic brakes on the first locomotive.[9]

At milepost 210.6 (13 miles [21 km] down the hill), the train began coming apart at 59 mph (95 km/h). The first 20 cars broke off, derailing 17; another 18 broke off and derailed at milepost 209.8. At milepost 208.2, 41 of the remaining cars derailed, destroying a house and killing a 15-year-old boy. The 3 locomotives finally came to a stop at milepost 206.5.

The investigators concluded that there was nothing the engineer could have done given the fact that two locomotives were operating without dynamic brakes. Also, the excessive operating speeds used on this hill do not allow any margin for dynamic brake failure. The engineer was faulted for a lack of situational awareness. He might have noticed that the train seemed to require excess use of the airbrakes.

New Rules

Over the years trains became heavier, but operating speeds remained the same on many dangerous hills. The railroads were becoming too reliant on dynamic brakes. Unfortunately, like all electronic equipment, dynamic brakes are subject to unexpected failures. Railroads are now required to operate at speeds that allow trains to safely stop without using dynamic brakes.

Other than how the train responds, the train crew cannot tell if the dynamic brakes are operating. The dynamic brake amp meter gives a reading, but only for the occupied locomotive. All new and rebuilt locomotives ordered after 2006 must now display a readout of dynamic braking forces for all locomotives and must be able to verify the DB are operating on all locomotives.

Frictional Power

Each brake shoe acts like a braking heat engine. The design target to prevent overheating is a maximum of 30 braking horsepower per wheel, which converts into 76,400 BTU/hour (22 kw)—about twice the heat output by all four burners on a typical kitchen gas stovetop.

Recall that the gravity force pulling on the last runaway train described (January 2000 in the Allegheny Mountains) on a 2.4% grade equals $0.024 \times 11,170$ tons $= 268$ tons, or 536,000 lbs (2,384 kN). The 3 dynamic brakes in the locomotives were supposed to provide 96,000, 81,000, and 60,000 lbs (427, 360, and 267 kN) of braking force. Since only the first locomotive had working dynamic brakes, 141,000 lbs (627 kN) of DB were missing; worse still, the crew was unaware.

To cancel gravity the required braking force from the airbrakes $=$ 536,000 lbs $-$ 96,000 lbs (from the working dynamic brakes) $= 440,000$ lbs (1,957 kN). If each wheel should not exceed 30 horsepower (22 kw) of braking power, the 80 cars with 8 wheels each have a total braking horsepower capacity of:

$$\frac{30 \text{ horsepower}}{\text{wheel}} \times 80 \text{ cars} \times \frac{8 \text{ wheels}}{\text{car}} = \frac{19,200 \text{ braking}}{\text{horsepower (14,317 kw)}}$$

Braking forces, torque, and horsepower are discussed in Chapter 9. Recall that power = force × speed. If the maximum braking horsepower is limited to 19,200 horsepower (14,317 kw) and the gravity force is fixed at 536,000 lbs (2,384 kN), the speed of the train descending the hill must be reduced to match the available horsepower. A braking power of 19,200 horsepower (14,317 kw) is equivalent to a 440,000 lbs (1,957 kN) braking force at 13.43 mph (21.6 km/h). The train was safe to descend with only one working DB and airbrakes at 13.43 mph (21.6 km/h).[10]

The brakes can safely produce more than 30 horsepower (22.3 kw) for brief periods of time. Testing after the accident concluded that the brakes were asked to deliver 50 horsepower (37 kw) per wheel at 20 mph (32 km/h) and 64 horsepower (47.7 kw) at 30 mph (48 km/h).

After the accident, the braking sequence captured by the event recorder was duplicated on a test rig that records brake shoe force, average wheel temperature, retarding force, coefficient of friction, and braking horsepower. This testing was done specifically to study brake fade.

The operating rules of this company permit power-braking (braking while applying the throttle) with restrictions. The rules allowed power-braking for up to 2 minutes when the brakes are reduced 18 lbs or more. This train braked at 2 minutes with a 17-lb reduction. The investigators concluded that the brakes overheated during power-braking just 3.6 miles (5.8 km) down the hill.

At that point during the simulation on the test rig, wheels were moving at 23.8 mph (38 km/h). The investigators measured an average wheel temperature of 431°F (222°C), a brake shoe force of 1,013 lbs (5 kN), a coefficient of friction of 0.434, and a braking horsepower of 64 (47.7 kw). Within the next 3.67 miles, the test rig measured the wheel temperature rising to nearly 600°F (315°C) and the coefficient of friction, the brake shoe force, and braking horsepower decreasing by 40%, 34%, and 22% respectively. At this point the engineer increased braking and accelerated brake fade.

The industry standard for brake shoes requires a braking force slowing the train of at least 400 lbs (1.8 kN) when pressing on one brake shoe with a force of 1,450 lbs (6.4 kN) at 20 mph (32 km/h) for 45 minutes. This is equivalent to a brake horsepower of 21.3 (15.9 kw) at a temperature of 600°F (315°C) and a somewhat degraded coefficient of friction of

0.28. A new standard being phased in requires a 600-lb (2.7-kN) retarding braking force when applying a brake shoe force of 2,250 lbs (10 kN) for 45 minutes at a speed of 20 mph (32 km/h)—equivalent to 32 braking horsepower (23.8 kw). The new standard requires improved brake shoe materials.

Electronically Controlled Pneumatic Brakes

Electronically controlled pneumatic brakes replace the air pressure–controlled triple valve with electronically controlled valves. With ECP brakes, the brake pipe, air reservoir, and airbrake cylinder remain the same. The triple valve is replaced by two electronically controlled valves controlling air flow between the brake pipe, reservoir, and brake cylinder. Braking occurs uniformly at the speed of light instead of at less than the speed of sound. Uniform braking reduces stopping distances by 40% to 60% and brake shoe wear by 20% to 25%.

The Westinghouse Air Brake Company argued against using electrically controlled pneumatic brakes in very long freight trains as recently as 1981, even though electrically controlled brakes have been used since the 1930s on shorter passenger trains. The voltage at the back of the train (per Ohm's Law, voltage = current × resistance) is simply not the same as at the front. The electromagnetically actuated valves would not work reliably when located randomly within long trains. If part of the train braked and part did not, train forces could rip the freight train apart. (See "Handling Long, Heavy Freight Trains" in Chapter 7.)

The voltage variation was solved with modern digital electronics and computer software. The microprocessor in the locomotive communicates to the microprocessors on each car. Upon startup, the control unit in the locomotive sends out an initialization signal to locate all electronically activated valves. As the cars respond to the startup signal, software automatically adjusts for variations in location and voltage. Digital signals are broadcast from the locomotive like a computer network connection. Each car's control unit can both send and receive data, allowing the locomotive to monitor the performance and health of the brakes in each car.

The problems with electrical connection described earlier remained for many years. In fact, until very recently wireless radio systems (without connections) were seriously considered. After years of testing, the problem has finally been solved with exacting mechanical and electrical requirements.

For example, the connector pull-apart forces must be between 100 and 400 lbs (0.44 and 1.77 kN) (this also minimizes vandalism), and the connectors must be assembled by hand without tools. Both requirements must be met after 1,000 assembly test cycles and at temperature extremes of −50°F (or −45°C coated with ice) and 150°F (65°C). To protect against arcing damage, the connectors must survive almost 10 times the normal voltage for 5 minutes.[11]

After many years of development and testing, ECP brakes were finally used in full time revenue service for the first time in the United States in 2007.[12]

There are many other advantages. Electronically controlled airbrakes allow gradual release and greatly simplified braking strategy. With ECP brakes, the engineer could have avoided applying the throttle (to avoid stalling) while braking downhill during the January 2000 coal train runaway. Instead, he would have gradually released the brakes and avoided further overloading the brakes by power-braking.

The electronic valves also allow continuous recharging of the reservoirs, making it virtually impossible to run out of air. ECP brakes save fuel and reduce trip time. Without gradual release, the tendency is to overbrake. The train moves slower than necessary and uses more fuel to accelerate back up to speed. Projections are a 5% fuel savings and a 10% reduction of trip time. Without gradual release, fuel is also wasted by occasionally using power-braking.

Also, there are less train stops required for brake inspection. Because the electric wire provides constant diagnostics on all the brakes, the Federal Railroad Administration (FRA) has extended the required 1,000-mile (1,600-km) brake inspection to 3,500 miles (5,632 km) for ECP brakes.[13]

UNIFORM BRAKING ALSO GREATLY REDUCES SLACK ACTION. Slack action, explained at length in Chapter 7, occurs because of movement between cars created by the draft gear that absorbs shock. Different parts of the train move at different speeds with forces violent enough to damage freight or even derail the train.

UDE, or undesired emergency braking, will be virtually eliminated. One railroad reports a UDE about every 4,700 miles (7,564 km) on ordinary trains as compared to none on their ECP test train. The traditional airbrake control valve became more complicated, incorporating added parts and ports to speed up brake response. The improved hair-trigger response became susceptible to tiny defects, perhaps a piece of dirt, which could set off emergency brakes in all cars. Also, ECP brakes are very important to the efficient application of Positive Train Control (see Chapter 6).

Barriers to ECP Brakes' Adoption

The same problems adopting Westinghouse's airbrake in freight cars are occurring with ECP brakes. The initial cost of converting the nation's 1.3 million freight cars is formidable. Most of the savings associated with ECP brakes will benefit the train operators. The private owners of half the nation's freight cars couldn't care less about fuel savings, quicker turnaround time, and decreased derailments.

There are two different ECP configurations: stand-alone and overlay. The overlay system works on top of conventional airbrakes. Freight cars with an overlay can operate in either ECP or conventional mode. Railroads have to commit to one or the other and develop new handling rules for trains with ECP. Just as 130 years ago, standardization of design must be agreed to for free interchange of cars. Sorting out cars with and without ECP will be problematic during the transition period.

The first trains being adopted for ECP use are unit trains, or trains that stay together for one purpose, such as a 100-car coal train that only moves back and forth between the coal mine and a power station. The railroad owns the cars, operates the train, and receives all the benefits of conversion. About one-sixth of the 1.3 million freight cars operate as unit trains.

PASSENGER TRAIN BRAKES

Because the nation's 1.3 million freight cars must connect interchangeably, all freight train brakes are more or less the same with only minor variations. Passenger trains, on the other hand, are owned and operated by one company and custom designed with a variety of braking configurations. Passenger brakes require redundant systems that are extremely reliable. Brake failure has not been the primary cause of a passenger train fatal collision in the United States in the modern era.

Passenger train brakes were identical to those of freight cars until 1905, when the first gradual release brake was introduced for passenger service. Gradual release provides better control and a more comfortable ride. Gradual release also requires far more complicated pneumatic logic than the freight car's triple valves.

The next development in passenger brakes was the use of electro-pneumatic (EP) brakes, first used on electric transit cars in the early part of the twentieth century. EP brakes, with the advantage of instant response and gradual release, use electromagnetic valves to control the airflow. EP brakes, not to be confused with digital electronically controlled pneumatic brakes, had all of the problems previously described with electrical control of brakes. Electric control was not considered reliable and was always backed up by normal airbrakes.

Passenger trains accelerate and decelerate at higher rates than freight trains. (People are more impatient than a pile of coal; also, passenger trains can accelerate more because of less mass.) Eventually, heat damage to the wheels and brake shoes limits braking operations. In the 1930s, disc brakes came into use on the high-speed streamliners.

Disc brakes add additional surfaces to dissipate the frictional heat. A disc brake is typically two discs separated by cooling fins. Calipers press brake pads against both discs. The faster the train moves, the more discs used. Shown in Figure 10.8 are three disc brakes mounted on an axle, similar to Amtrak's 150-mph (241-km/h) Acela design.

Anti–wheel slip systems were developed in the 1940s to minimize overbraking and wheel slide (and wheel damage). Early mechanical schemes were replaced by electrical systems with small generators attached to each axle. Any difference in generated voltage indicated wheel

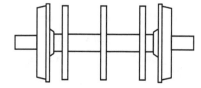

FIGURE 10.8. Axle with two wheels and three disc brakes mounted on the axle.

slip and triggered reduced braking. Today passenger brakes are computer controlled and are similar to a car's anti-lock brakes.

Blended brakes, combining the locomotive's dynamic brakes with airbrakes, were developed for passenger trains in the 1970s. Again, better control and a smoother ride were the goals. Amfleet cars, used by Amtrak on the Northeast Corridor since the 1970s at speeds up to 125 mph (201 km/h), use blended braking, two disc brakes on each axle, a wheel-slip protection system, and conventional airbrakes. Amfleet's disc brakes supply 60% of the braking effort.

Acela's Brake Problem

On April 14, 2005, Amtrak's 150-mph (241-km/h) Acela had just completed 4 days of mandated annual testing. After the test runs, FRA inspectors discovered a metal fatigue crack on one of the disc brake spokes. Inspections of all 1,440 discs on all 20 Acela train sets discovered 300 cracked brake discs. In some cases individual spokes had completely cracked. All Acela traffic was stopped on the morning of April 15.

Although the brake manufacturer recommended inspection of the brakes every 12,400 miles (20,000 km), somehow that requirement never got into the final Amtrak maintenance manuals. A year earlier, a subcontractor doing maintenance on the wheels reported finding cracks in the disc spokes. In yet another communication failure, Amtrak was not notified.

Nevertheless, no discs broke off and no accidents occurred. An aggressive engineering testing and analysis program returned the trains to full service on July 11, 2005. Fortunately, a replacement disc design already existed. Early in the investigation it was discovered that the original brake design, with heavier, stronger spokes, had been replaced by a different design with lighter, weaker spokes.

The problem proved to be unexpected vibrations of the disc during some braking operations. The vibrations did not occur on every test. The vibrations, when they did occur, grew with increased brake pressures. The vibrations added to the thermal expansion stresses during normal braking and fatigued the spokes. Fortunately, the older design with stronger spokes did not vibrate.

With 72 discs on each train (plus dynamic brakes), one broken disc posed little braking hazard. If all spokes fractured, however, a loose disc could dangle around the axle at 150 mph "like a bracelet on a broomstick."[14] This could damage the axle and derail the train. A wheel broken off would create substantial noise and vibration. Although an unacceptable risk, the train crew could hopefully respond quickly enough to safely stop the train.

11

Broken Rail

HIGH-SPEED TRAIN DERAILS IN ENGLAND

On October 17, 2000, a passenger train departed London with 182 passengers and crew. The train consisted of a locomotive, 8 passenger cars, a restaurant car, and a cab car. The passengers in front heard a loud bang as the steel wheels went off the rails at 116 mph (187 km/h) near Hatfield, England. The locomotive engineer (called a driver in England) thought he had hit something on the tracks. About 115 feet (35 m) of track on the outside rail of a sweeping curve (radius 4,790 feet [1,460 m]) had disintegrated into hundreds of fragments. The inside rail, repeatedly gouged by derailing cars, remained intact.

Even without an event recorder, the investigators were able to piece together the accident sequence by studying track gouges, passenger car damage, roadbed damage, and final car locations. The first rail fracture could be identified with reasonable certainty—the site with the largest pre-existing metal fatigue crack and the site with the most wheel gouges.

The locomotive and first two cars passed over the rapidly fragmenting rail. Cars 3 through 6 were derailed by the damaged track. The broken rail was now greatly displaced and fragmented, and flipped Car 7 on its side. The train separated behind overturning Car 7. The separation automatically triggered the emergency airbrakes. Car 8, with 2 cars still attached, dug into the gravel roadbed and rapidly came to a stop 680 feet (207 m) past the initial rail break. The locomotive and attached 7 cars continued for nearly 2,100 feet (640 m)—dragging overturned Car 7.

Car 7 began sliding on its side about 190 feet (58 m) past the first rail break and was dragged almost 500 feet (152 m) before striking a utility pole. The pole was torn from its concrete base and partially sheared off the car's roof. Car 7 continued sliding another 380 feet (116 m) and struck a second utility pole. The second pole crushed the roof down to the floor. Car 7 finally stopped almost 600 feet (183 m) after striking the second pole. All 4 fatalities and 2 of the 4 serious injuries were in Car 7.

A survivor from the seventh car tells her story. "I felt quite silly as I thought I had just fallen over, I must have hit my head. Then the train lurched down violently and turned over. There was a ripping sound. It was very misty and then a window blew out. Things slid on my head. There was dust everywhere. I remember looking at the track and it was coming very fast beneath me. I kept thinking about my little boy and that I might not see him again and if I didn't hold on to this bar I would get sucked out of the train. I had a mental picture of my boy's face. I thought I had just got to hold on—hold on. I had dust in my eyes. The train came to a stop. It was completely quiet. I could see a pile of bodies on top of me. I wiggled my fingers and toes and I thought I wasn't paralyzed."[1]

Over the next 4 days, fingertip searches recovered about 200 rail fragments (almost 90%) of the shattered 115 feet (35 m) of rail. It is unusual for this much fragmentation. The rail was later found to have nearly 50 pre-existing metal fatigue cracks—the proverbial accident waiting to happen. More fragmentation occurred as the wheels impacted the twisted rail at 116 mph (187 km/h).

The track at Hatfield had been replaced in 1982 and again in 1995. Because of increased traffic, it was decided in February 2000 to replace the rails again. Plans to reroute or cancel trains were made and a 27-hour window in March was scheduled for the work. The replacement rail did not arrive until April. Track possession was harder to negotiate during the busier spring schedule. Railtrack (the owner of the track) offered two eight-hour time slots; the contractor said they needed five. Track replacement was rescheduled during November, when there would be less traffic. The train derailed in October on top of the new replacement rail lying along the track.

There were many serious problems with Railtrack's maintenance procedures. For example, how can track be visually inspected when a 116-mph (187-km/h) train approaches with a 4-second warning? Inspectors reported that there had been fatigue damage without any follow-up repair, or that rail damage went unnoticed because of faulty procedures or poor training.

Railtrack knew that it had problems with track repair. Rail breaks had increased to excessive levels. There were numerous letters documenting the problems before the disaster. One contractor had been fired for poor performance. Without question the rail had extensive metal fatigue damage requiring immediate attention—at least a speed reduction until repairs could be made. Several officials were accused of manslaughter. Eventually, individual charges were dropped. Railtrack and its contractors were heavily fined.

How was the track allowed to degrade? There was a backlog of hundreds of inspection reports and a similar list of repairs. Lacking the expertise to effectively prioritize repair work, Railtrack simply lost control of the situation. Urgent track repair orders were lost in the system. Railtrack got behind in repair work and never caught up.

For our purposes, the Hatfield accident also triggered numerous technical studies of metal fatigue in rails.

RAIL: HISTORICAL PERSPECTIVE

The cross sectional shape of rail has gone through a variety of changes over the years. The shape used today (Figure 11.1) is quite complex, with many transitions and radiuses that optimize mechanical stresses and manufacturing requirements. It's common practice to describe rail as lbs/yd, with a higher number indicating a heavier, stronger rail.

In the United States in 1920, almost 20% of new rail was 45 lbs/yd (22.3 kg/m) or less. By 1940, nearly half of all existing rail was heavier than 100 lbs/yd (45 kg/m). Today the most common rail in the United

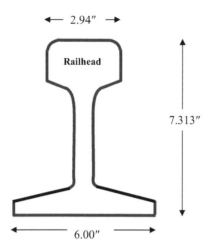

← 2.94″ →

Railhead

7.313″

6.00″

FIGURE 11.1. Cross section for rail rated at 136 lbs/yd.

States is 115 lbs/yd (57 kg/m) on low-density track and 132–136 lbs/yd (65.5–67.5 kg/m) on heavy haul main lines. The rail used at Hatfield, England, was 113 lbs/yd (56 kg/m).

Rail became slowly heavier and stronger as freight cars did the same. The average freight car capacity in the beginning of the twentieth century was 25 tons (22.7 MT). By the mid-1990s, the 115-ton (104-MT) capacity car weighing 286,000 lbs (129,700 kg) gross came into widespread use. In spite of continuous economic incentive to haul heavier cars, change occurs slowly. With heavier cars everything else must eventually be upgraded, including track, ties, roadbeds, bridges, wheels, axles, and bearings.

The rail shown in Figure 11.1 is 136 lbs/yd (67.5 kg/m). Lighter rail is still common on secondary track. A 2001 study of secondary track in North Dakota showed 230 miles (370 km) of 66, 68, 70, and 72 lbs/yd rail being used and almost 1,000 miles (1,609 km) of 75, 77, 80, and 85 lbs/yd rail. Most of it is 50 years old or older.

FRA Track Classes

In 1970, the Federal Railroad Administration (FRA) established for the first time classes of track with different speed limits, as shown in Table 11.1. Because of their lighter weight and higher safety standards, passenger trains are permitted to operate at higher speeds. The lighter passenger cars are less likely to damage the rails, axles, bearings, and

TABLE 11.1
U.S. Railroad Track Classes

| Class | Speed Limits (mph) | |
	Freight	Passenger
1	10	15
2	25	30
3	40	60
4	60	80
5	80	90
6	110	
7	125	
8	160	

wheels. With more frequent stops, passenger cars are less likely to have unnoticed damage from normal wear and tear. The people onboard can often hear problems developing (i.e., noisy bearings and wheels, dragging equipment, etc.).

Most of the main lines of the largest railroads are Class 4 or 5. Only Amtrak's Northeast Corridor is Class 7 or 8. Higher classes of track have heavier track structure, are more frequently inspected, have better maintenance, and have tighter tolerances on misalignments.

Wear

Today rail typically makes up 60% of a railroad's assets, and companies spend a significant part of their budget maintaining rail. In 2009, Union Pacific spent more than $1.7 billion maintaining its 50,885 miles (81,890 km) of track (including yards and sidings).

The most common reason for replacing rail is wear. More than 500,000 tons (453,600 MT) of replacement rail (about 4,000 miles [6,437 km] of rail) is installed each year in the United States.

Wear occurs from the steel wheel rolling on the rail. Under load the contact point distorts into an area about the size of a dime. This contact area disrupts the perfect rolling motion of the wheel, causing a small sliding or rubbing that causes wear. Contact wear can be seen as a polished shiny region on the top of the rail. On straight track, wear is on top of the rail. On curved track, the wheel flange also rubs the side of the rail

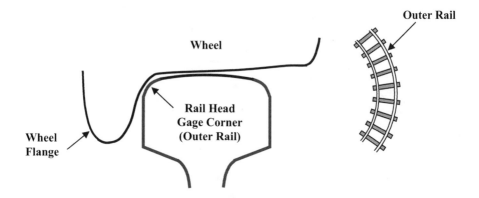

FIGURE 11.2. On curves the inside corner of the outer rail wears the most and potentially grows fatigue cracks.

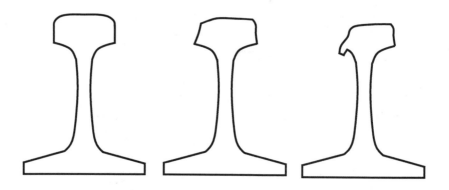

FIGURE 11.3. Wear on a curve.

(Figure 11.2). Example wear on the outside rail of a curve is shown in Figure 11.3.

Sometimes lubrication is used to decrease the side rubbing and wear on curves. Soft steel wears faster than hard steel. For that reason steel has continued to evolve into a harder, stronger, more wear-resistant material. Harder steels, lubrication, and/or trucks with better steering around curves all reduce wear.

RAIL USAGE IS OFTEN DESCRIBED in terms of gross million tons (MGT) of traffic. (Assuming 10,000 tons per train, 100 MGT occurs after 10,000 trains.) MGT is used to set rail inspection and replacement schedules. A high-traffic line might be 50 MGT per year or greater. One particularly heavily used track hauled 445 MGT of coal per year to a power plant. Worn rail will eventually have excessive failure rates or other stability problems.

A recent survey showed average replacement on a sharp curve (radius 575 feet [175 m]) after 420 MGT and 1,500 MGT for straight track (commonly called tangent track). Larger-radius curves will wear at in-between rates.

Different companies will set different limits on rail wear based on their experience and operating conditions. Example limits used by one company are 0.8125 inch (2 cm) on top and 0.625 inch (1.58 cm) on the side. When these limits are exceeded, trains are slowed until the rail is replaced.

Broken Rail

Most rails wear out before they break. A broken rail, however, is more likely to derail a train. The two are related. A worn rail has a higher chance of fracture because the stresses increase as metal is worn away. Numerous partial rail breaks found during inspection will often trigger a faster replacement schedule. One example replacement schedule replaces the rail if two to three metal fatigue defects occur within 20 MGT of traffic.

The major source of broken rail is metal fatigue. Metal fatigue occurs from contact between the wheel and rail and from rail bending. Metal fatigue can be explained as simply as bending a paperclip back and forth until it breaks, analogous to the minute rail bending that occurs with each passing wheel. Just as rail gradually deteriorates from wear, it also gradually deteriorates from metal fatigue.

The strength of a material, called the tensile strength, can be determined with a single one-time breaking load. Modern rail steel has a tensile strength of about 160,000 psi (1,100 MPa). This means a weight of 160,000 lbs (72,570 kg) hanging from a bar with a 1 inch^2 (6.45 cm^2) cross sectional area (6.45 cm^2) is about to break.

A metal's strength can also be determined for repeated cyclic loading. The breaking load for repeated loading is called the material's fatigue strength. The fatigue strength can be considerably less than the tensile strength because repeated loading causes damage on an atomic scale. Higher rail forces lead to lower fatigue life. This is analogous to bending the paper clip with a larger angle to cause fracture with fewer bending cycles.

Given identical test samples, the tensile strength is quite repeatable. The fatigue strength, on the other hand, has tremendous statistic scatter that depends on many factors.

How long does it take to break a rail with metal fatigue? It is a two-step process. First the fatigue crack must form in an incubation process commonly called crack initiation. The fatigue crack must then grow, or "propagate," until the rail is too weak to carry the wheel load. At that point the rail fractures. Inspection strategies that minimize rail breaks require an understanding of how long it takes to incubate and grow a fatigue crack.

The size of the fatigue crack growing in the rail head is normally described as a percent of rail head cross section that is cracked (i.e., disconnected or separated). Although the inspection technology keeps improving, defects smaller than 10% of the rail head are normally considered too small to find reliably. One rule of thumb considers fracture imminent for an 80% cracked head. The period of growth between 10% and 80% of cracked rail head is considered a window of opportunity to find the rail defect by inspection.

Fatigue Crack Incubation and Growth

How long do fatigue cracks take to incubate? The required rail traffic to incubate a crack is highly statistical and depends on many conditions. Nevertheless, studies have been done to determine ballpark answers. Railroads will have their own data for their operating conditions.

One study established statistical equations that predicted fatigue crack incubation as a function of track use. Their prediction for 100 miles (161 km) of track after 35, 50, 70, and 100 million gross tons (MGT) of traffic respectively is that 1, 2, 4, and 8 fatigue defects will grow within the next 10 MGT of use. In other words, 100 miles of track used for 35 MGT will grow on average 1 defect within the next 10 MGT of use. If the 100 miles of track was used for 50 MGT it would be expected to grow 2 defects within the next 10 MGT of use. As with all statistics, these values are

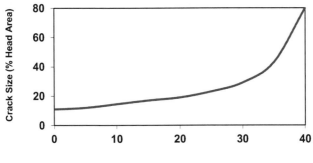

FIGURE 11.4. Crack growth data after crack incubation. Actual data plotted on this curve would show tremendous scatter. *Adapted from Palese 2000.*

realistic projections; actual defects formed will vary widely. One funny thing about statistics (and there are a few) is that some small percentage of track will incubate fatigue cracks after much less traffic.

After the crack incubates, how long does it take to grow to a critical length? Average crack growth rates estimated from another study are shown in Figure 11.4. The total life (MGT) until rail fracture equals the incubation period plus the crack growth period.

PREDICTING HOW LONG a fatigue crack takes to incubate, how fast the crack grows, and what wheel load ultimately fractures the rail is a bit like predicting the weather. The incubation, growth, and fracture data all have great statistical scatter, and different researchers get different results.

Also, statistical predictions become more reliable for a large group as a whole. Given 100,000 people, we can reasonably predict what percentage will die in the next year even though we have very little ability to predict death for individuals picked at random. To continue the analogy, specific rail (or people) can be identified by inspection for more careful monitoring.

RAIL STRESSES

Wheel loads result in two types of mechanical stresses in the rail: bending and contact stress.

First recall the meaning of stress: load per unit area. A steel rod with a cross sectional area of 13.35 in² (86 cm²) (the same as 136 lbs/yd rail) hung from the ceiling and supporting 100,000 lbs (45,360 kg) has a tensile stress of 100,000 lbs divided by 13.35 in² = 7,490 lbs/in², commonly referred to as 7,490 psi (51,640 kPa), the same units as pressure.

A typical example of modern rail steel has a yield stress (the stress that will cause permanent deformation) of 90,000 psi (620 MPa) and a tensile breaking stress of about 160,000 psi (1,100 MPa). Showing how remarkably strong steel rails are, the bar above would not be expected to break until a load of 13.35 in² × 160,000 lbs/in² = 2.1 million lbs (969 MGT), the weight of about 611 cars—a staggering force!

Bending Stresses

Bending stresses are not uniform like the tensile stresses described earlier in this chapter but vary across the height of the cross section.

Consider a steel bar clamped onto a table top and loaded, as shown in Figure 11.5. The bending in the bar is exaggerated to show how the top of the bar elongates in tension and the bottom of the bar shortens in compression. Cutting the bent bar with a saw also illustrates the difference between tension and compressive stresses. If the saw cut is made on top, the tensile stresses open up the saw cut. If the saw cut is made on the bottom, the compressive stresses squeeze the saw blade, making it difficult to saw—just like sawing a bent tree limb. Repeatedly loading the end of the bar causes a fatigue crack on the top surface, as shown.

The rail sits on an elastic or flexible foundation. The ties, embedded in crushed rock ballast and supported by soil, do in fact flex a bit. Rail engineers describe this flexibility as if the steel rail sat on extremely strong springs spaced at the tie spacing, as shown in Figure 11.6.

When the wheel presses down on the rail, the push-back from the springs is distributed over many springs, as shown in Figure 11.7. The distributed push-back from the ties causes rail uplift in front of and behind the wheel. Because of this bending pattern, tensile bending stresses exist on the top of the rail in front of and behind the wheel and on the bottom of the rail directly under the wheel.

The bending stresses are highest on the bottom rail surface directly underneath the wheel. Fatigue cracks, however, have a strong tendency

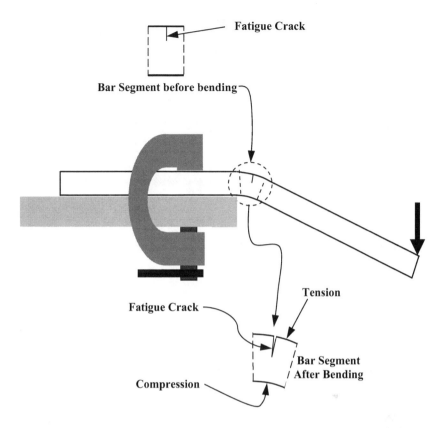

FIGURE 11.5. Bending stress in a clamped bar hanging over the edge of a table. Note the tension on the top of the bar and the compression on the bottom.

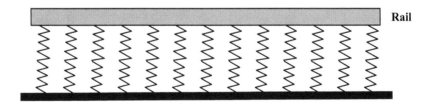

FIGURE 11.6. The ties act like springs that provide flexible support.

to grow on the top surface, where the bending stresses interact with the contact stresses directly underneath the wheel.

Track deflection under the wheels is also used as an overall indicator of the condition of the rail-tie-ballast "system." Excess deflection is a bad

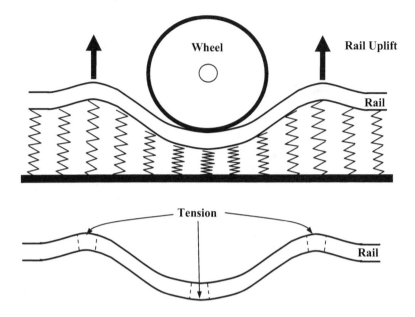

FIGURE 11.7. The flexibility of the springs creates uplift in the rail (greatly exaggerated). The rail curvature creates tensile bending, as shown.

thing. The rule of thumb is that track with deflections less than 0.2 inch (0.5 cm) should last indefinitely and that deflections greater than 0.4 inch (1 cm) will deteriorate quickly. Excess deflection will wear out the ties, crushed rock ballast, and underlying soil and cause even greater deflections that overstress the rail. The ballast can break down, abrade the rock, and affect drainage. Eventually, individual ties will "pump," or move up and down several inches, in a mud hole with each passing wheel.

Contact Stresses

The weight of each car is supported by eight wheels. All weight is focused on eight tiny spots about the size of a dime. The contact force divided by the contact area creates a contact stress of about 200,000 psi (1,379 MPa) for well-shaped wheels and rail.

Rail-bending fatigue is different than surface contact fatigue. For most metal fatigue (including rail-bending), the fatigue crack grows when the crack opens up and extends, as shown in Figure 11.8.

The contact stresses directly under the wheel are compressive. Nor-

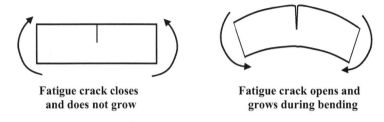

**Fatigue crack closes
and does not grow**

**Fatigue crack opens and
grows during bending**

FIGURE 11.8. Most fatigue cracks grow when the crack opens up.

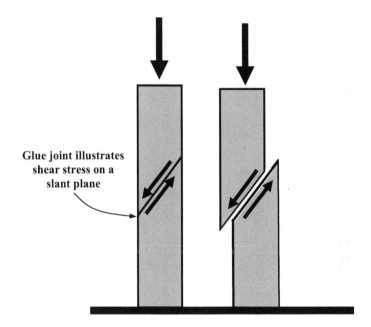

**Glue joint illustrates
shear stress on a
slant plane**

FIGURE 11.9. Shearing or sliding stresses can initiate a fatigue crack, even
though the bar is loaded in compression.

mally, fatigue cracks do not grow with compressive stresses because the
cracks do not open up. The stress state under the wheel, however, is quite
complex. Fatigue cracks associated with contact will occur and grow with
a sliding or shear stress below the surface.[2] Figure 11.9 illustrates how a
bar loaded in compression will in fact have shear or sliding stresses. A
steel bar is sliced at a 45-degree angle and glued back together. Clearly,
the bar will fail at the glued joint with shearing stresses if the compressive force is large enough. Slightly below the contact footprint of the

wheel, the stress patterns create shearing stresses that can grow a fatigue crack.

Contact stresses between the rail and wheel are localized on the surface and quickly dissipate below a depth of 0.5 inch (1.3 cm) or so. In fact, the vast majority of surface fatigue cracking penetrates to near that depth and stops.

Rolling contact fatigue cracks may appear as fine cracks on the surface. The cracks grow at a very shallow angle, 5 to 15 degrees, relative to the horizontal surface. Spalling occurs if multiple cracks join and metal falls out. Depending on the friction at the wheel-rail interface, the maximum shearing stress can be very close to the surface. Lower surface friction creates deeper contact fatigue damage.

A new type of subsurface rolling contact fatigue defect began appearing in the 1940s under 50-ton (45-MT) and 70-ton (64-MT) cars on curves. This defect, known as "shelling," appears as a dark streak on the rail's top surface. The subsurface metal disintegrates and flows sideways, creating a depression on the surface (the shell). As the rest of the running surface is polished shiny with wear, the depressed shell stands out as a dark spot.

With the introduction of 263,000-lb (119,000-kg) cars in the United States during the 1960s, shelling began appearing on straight track. Shelling is a subsurface defect that grows parallel to the top of the rail surface. Shells that remain parallel are considered harmless unless they turn and grow down into the rail head in the transverse direction (Figure 11.10). If the crack turns down into the rail head, further crack growth is controlled by rail head bending stresses; growth continues until the rail fractures. In the Hatfield, England, derailment there were surface cracks, spalling, and heavy shelling on the rails.

The worst possible scenario is several transverse fatigue cracks growing in close proximity. If all the cracks break at once, the rail shatters— the problem at Hatfield. (Recall that 115 feet [35 m] of rail at Hatfield shattered into nearly 200 fragments.)

The inside corner of the outer rail on curves is particularly susceptible (see Figure 11.2). In this case the defect is called gage corner cracking, and after sufficient growth it will become a detail fracture. Multiple detail fractures occurred at Hatfield.

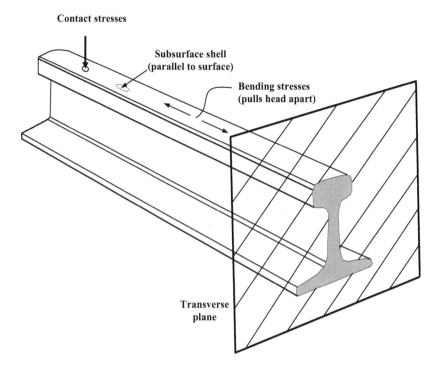

FIGURE 11.10. Orientation of contact stresses, subsurface shell, bending stresses, and transverse plane.

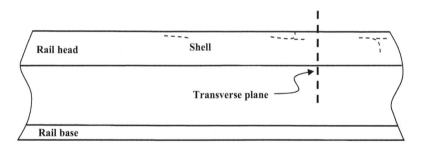

FIGURE 11.11. Side view of a rail with three subsurface rolling contact fatigue defects. One crack remains very shallow, the other two branch up or down.

Contact fatigue is not inevitable. If the rate of wear exceeds the growth of surface contact fatigue cracks, contact fatigue damage can be worn away as fast as it develops. If the wear rates are low, however, the contact fatigue cracks are more likely to grow into rail fractures.

Railroad companies will often grind the surface to remove surface defects, maintain an optimum profile for wheel contact, and clean up the surface for ultrasonic inspection. In 2006, Canadian National ground more than 16,000 miles (25,750 km) of its 20,264 miles (32,600 km) of track.

Typically, 0.002 to 0.008 inch (0.005 to 0.02 cm) of surface is removed as a form of artificial wear. Rail engineers sometimes speak of the "magic" wear rate that removes surface contact fatigue cracks as fast as they appear. Ideally, a grinding program aims to artificially create the magic wear rate. Grinding frequency depends on MGT, axle loads, speed, wheel contours, locations of frequent braking and accelerations, and the hardness of the steel rail.

At Hatfield there were 5 or 6 grinding passes removing about 0.008 inch per pass before the derailment. The spalling and surface cracking, however, exceeded these depths. The grinding did not remove the surface cracks.

Wear is still the *dominant* rail problem that drives rail replacement rates. As freight cars become heavier and newer steels become more wear resistant, metal fatigue problems are trending up.

OTHER STRESSES IN THE RAIL

In addition to bending stresses and contact stresses, there are thermal stresses and residual stresses. Residual stresses are left over from the manufacturing process and occur with no external loading on the rail. The high-contact stresses between the wheel and rail also create a thin layer of residual stresses on the running surface. The rail thermally expands and contracts with the ambient weather, creating thermal stresses. Thermal stresses can cause track buckles in the summer and rail breaks in the winter (see Chapter 12).

Wheel Impact Loads

Imagine a faceted bowling ball trying to roll down the alley— bam, bam, bam every rotation. A wheel with a flat spot will impact the rail with every rotation. Frequently, a rhythmic banging can be heard on a passing train because of the impact of a minor flat spot. Flat spots occur when the brakes stick for a variety of reasons. The wheel slides instead of

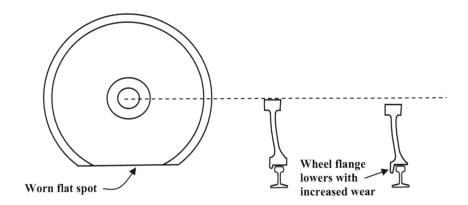

Worn flat spot

Wheel flange lowers with increased wear

FIGURE 11.12. Severely worn wheel resulting from wheel slide (shown approximately true scale).

rolling and a flat spot is quickly worn. A rail can break when a high wheel impact load randomly finds a rail with a growing fatigue crack.

A severe example of wheel wear occurred on a Canadian National passenger train on March 21, 1996. The brakes locked (water in the disc brake froze the actuator) and the wheel slid for 240 miles (386 km), creating a 15.5-inch (39-cm) flat spot on the wheel (Figure 11.12). The wheel kept wearing lower until the wheel flange caught on a railroad switch. The passenger train had just slowed to 30 mph (48 km/h) to enter a work zone. The track work crew saw a derailed wheel and radioed the engineer, who immediately applied the emergency brakes. Just one truck on one car derailed. All the cars remained upright. Although this was a relatively minor derailment with no injuries, the train had been going 91 mph (146 km/h) just minutes before on track authorized for 100 mph (161 km/h).

ON MARCH 12, 2001, a Canadian Pacific train consisting of 2 locomotives and 89 freight cars derailed 14 cars when a 114-inch (2.9-m) length of 115 lbs/yd rail shattered into 56 pieces. The train was traveling at 40 mph (64 km/h).

Two out-of-round wheels, with flat spots about 2.5 inches (6.35 cm) long, were found after the accident. One wheel, 0.115 inch (0.29 cm) out-of-round, had estimated wheel impact loads of 108,000–145,000 lbs (48,988–65,771 kg). The other wheel, 0.185 inch (0.47 cm) out-of-round,

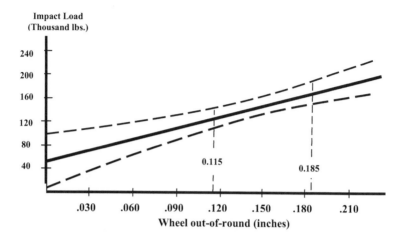

FIGURE 11.13. Wheel impact load versus wheel out-of-round data. *From TSB Report R01H0005.*

had impact loads of 151,000–191,000 lbs (68,492–86,636 kg) (Figure 11.13). Compare this to a fully loaded 283,000-lb (128,366-kg) hopper car supported on 8 wheels, or 283,000/8 = 35,375 lbs (15,875 kg) per wheel.

Wheels are condemnable based on a variety of geometric defects that are difficult to spot and time consuming to measure. Increasingly, companies are using wheel impact load detector (WILD) systems to detect bad wheels. There were 50 WILD systems in North America in 2002, 85 in 2007, and 131 as of 2010. Between 2004 and 2010, the number of wheel loads greater than 140,000 lbs (63,500 kg) has been reduced by 75%. WILD uses a series of strain gages on the rails to directly measure the wheel impact loads. If a high-impact wheel is detected, a message is automatically sent to the dispatcher or train crew.

Surface fatigue can occur on wheels or rails. The surface damage can be "healed" by plastic flow, creating out-of-round or oblong wheels. Agreed-upon rules condemn a wheel if out-of-round by 0.070 inch (0.18 cm) or if it has a WILD reading greater than 90,000 lbs (40,800 kg). The rules allow the operator of the tracks to repair a wheel with a WILD reading of 90,000 lbs or higher and charge the car's owner. A wheel load of 140,000 to 160,000 lbs (63,500 to 72,600 kg) will generally result in the train being stopped for immediate wheel repair.

Wheel impacts are not exactly a new problem. In 1912, 200 rail breaks were recorded in a 14-mile (22.5-km) stretch from a single passing train.

Strength of Fatigue-Cracked Rail

As the fatigue crack in the rail continues to grow, exactly when will the rail break? Limited data are available. Once again, the process is highly statistical.

Four 5-foot (1.5-m) sections of rail (132 lbs/yd) containing fatigue cracks of various sizes in the rail head were tested for strength. Each span of rail was supported on each end and loaded in the center above the fatigue crack (Figure 11.14).

This is not exactly the same as rail sitting on ties spaced every 20 inches (51 cm) or so, but is useful for comparison. The test results shown in Table 11.2 can be compared to the expected strength of rail without a fatigue crack—estimated to be at least 240,000 lbs (108,800 kg), the weight of about 68 automobiles. At this load the rail will bend excessively. Column 4 reinterprets the breaking load into an equivalent number of automobiles.

The fatigue cracks listed in Table 11.2 are seen to weaken the rail by 50% to 90% of the crack-free strength of 240,000 lbs. The relatively small crack of 3% of the head area reduces the load-carrying capacity by almost 50%.

Because fracture data always have tremendous statistical scatter, the loads shown are merely example data. Repeated testing of apparently identical specimens will show higher or lower breaking loads. The data do show the expected general trend that increased fatigue crack size lowers

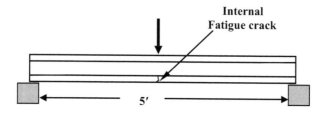

FIGURE 11.14. Strength test of rails with fatigue cracks.

TABLE 11.2
Breaking Load versus Fatigue Crack Size

Test	Size of Fatigue Crack (% of head area)	Breaking Load (lbs)	Number of Autos Supported (3,500 lbs per auto)
1	3	123,000	35
2	22	148,000	42
3	46	30,000	8.6
4	89	23,000	6.6

the strength of the rail. The trend, however, is not perfect. The second test shows a bigger fatigue crack supporting more load than the first.

Ultrasonic Rail Inspection

The first line of defense against metal fatigue is inspection. There are several different types of track inspection, including visual, geometric, and ultrasonic inspection.

Fatigue cracks growing inside the rail can be seen up close visually when they break through the surface. But usually surface cracks cannot be seen by the frequently passing visual inspectors. An inspector riding in a hi-rail truck is expected to feel and hear a completely broken rail but not a partial break. Internal cracks can only be seen ultrasonically in a process similar to that used by doctors.

Ultrasonic inspection can be defeated by a poor rail surface. If the surface is covered with surface spalling and fatigue cracks, the ultrasonic sensor will not get a clean reflection off the bottom of the rail.

THE ULTRASONIC INSPECTION at Hatfield, England, was required every six months. Unfortunately, with too many contractors and subcontractors there was confusion about adequate follow-up. In November 1999, the Hatfield ultrasonic inspector noted severe gage corner cracking and intermittent loss of the ultrasonic reflection off the rail bottom. The inspector reported total loss of the reflection off the rail bottom four months later, with no follow-up in between. And in June 2000, the inspector reported that the rail was untestable because of severe surface chipping.

Surface grinding finally occurred three years after it had first been ordered, just one month before the accident in September 2000. Unfortunately, the surface fatigue damage was deeper than the grinding. Eleven days before the accident, the inspector stated that the rail was still ultrasonically untestable. This should have raised a major red flag about the condition of the rail.

In the United States, ultrasonic inspection frequency is based on use and class of track. For passenger trains operating above 110 mph (177 km/h) (Class 6 and higher), the track must be ultrasonically inspected twice a year. If the deadline is missed, the train's speed must be reduced to 25 mph (40 km/h) until inspection is completed.

TYPICALLY, three different ultrasonic sensors take readings from three different angles. The signals are amplified, processed, and combined with computers to minimize errors. But errors still occur. Just as in medicine, the ultrasonic inspection of rails depends on the training and experience of the inspector and the hardware and software of the inspection equipment that is being used. Accuracy is also affected by rail surface condi-

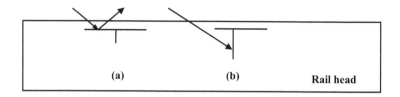

FIGURE 11.15. Case (a): The ultrasound signal (arrow) is reflected by the horizontal defect. The transverse defect is not found. Case (b): The ultrasound signal finds the transverse defect.

tions, dirt, and grease. Defect size and orientation also affect results. Surface and horizontal defects can mask a transverse defect from the ultrasonic signal, as shown in Figure 11.15.

The probability of finding a crack is greater for larger defects. One study by the Transportation Board of Canada states that there is a 68% chance of finding a 10% head area crack and an 88% chance of finding a 30% head area crack. Technological improvements continue to advance these numbers.

It takes many wheel cycles to incubate and grow a fatigue crack to a detectable size. The fatigue crack continues to grow until a wheel load exceeds the remaining strength of the partially cracked rail. The trick is to inspect often enough to find and repair the fatigue damage before it threatens train safety.

Numerous maintenance techniques can reduce rail fatigue, such as better management of out-of-round wheels, surface grinding of rails and wheels, maintenance of track geometry, and lubrication on curves. Railroad companies are also increasingly using "risk management" to optimize defect detection. Risk management considers frequency of crack incubation, expected fatigue crack growth rates, probability of defect detection, and track use to set inspection intervals. Increased track use in general calls for increased inspection. Inspection also increases when the risks increase, which means that tracks hauling passengers or hazardous material get more attention.

Probability of Train Derailment

Surprisingly, the most likely outcome of a broken rail is— nothing. In fact, operating rules actually permit a train to pass a broken rail at 10 mph (16 km/h) if an inspector is present at the broken rail and in radio contact with the train crew.

DATA FROM THE 1980S and more recent data collected from 1995 to 2003 indicated 770 and 826 rail fatigue cracks for every broken rail derailment respectively. Of these, 75% were found during routine ultrasonic inspection. The remaining 25% of the fatigue cracks were found during active train service. These "service defects" can be found in a variety of ways.

On signaled track a broken rail can be detected when the electrical current is interrupted. The electrical signal is only stopped for a completely broken rail. A broken rail can also be detected by the track patrol or by a passing train crew noticing rough track. The track patrol trucks typically run several times a week. Because they do not use ultrasonics, they can easily miss a crack unless the rail is completely broken. Hopefully, they will hear or feel a broken track. Only luck prevents a service defect from becoming a derailment. A great deal of luck, however, is normally expected. In a 2008 survey only 1.4% of service breaks derailed the train.

If a train derails from a broken rail that has been inspected ultrasonically, one of two things has occurred. The ultrasonic inspection failed to find the defect, or the fatigue crack defect grew since the last ultrasonic inspection.

WITH 770 TO 826 DEFECTS PER YEAR occurring for every broken track derailment and 250 to 300 broken track derailments per year in the United States, the total number of fatigue cracks occurring per year is estimated to be something like 200,000 to 250,000.

A single complete transverse rail fracture creates a potential separation of about an inch (2.54 cm) during cold weather and much smaller during warmer weather. The fracture and lack of track continuity results in a step change in track height. The discontinuity also increases stresses in nearby track. Wheels may batter the track into a serious misalignment that can derail the train, or the now weakened track may fracture at additional locations. Of course, the worst-case scenario is what happened at Hatfield. A high-speed passenger train broke a rail. The resulting impact load, misalignment, and multiple existing fatigue cracks shattered 115 feet (35 m) of rail and caused a serious derailment, an extremely rare event.

ELIMINATE BROKEN RAILS?

It's a bit like removing all potholes from the nation's roads. In 2004, more than 230,000 potholes were filled in New York City alone. Similar to cyclic fatigue loading in rail, the number of potholes is directly related to the number of freeze-thaw cycles.

Potholes would be far more dangerous if they grew internally and suddenly opened up on the surface like a broken rail. Consider the safety issues of driving along and suddenly a 12-inch (30-cm)-deep pothole opens up. Consider how difficult it would be for street crews to keep up if they had to search internally for potholes with special equipment instead of visually spotting them on the surface. For railroads the biggest problem is adequately monitoring 2 × 162,000 miles (260,714 kg) of rail and the 10.8 million wheels on the 1.35 million freight cars.

ALSO, THERE IS THE ISSUE OF ACCEPTABLE RISK. A freight train derails in the middle of nowhere, and there is limited risk to human safety. The same freight train slows down to enter a city and the risk of derailing at slower speeds is reduced. Passenger trains require higher standards. The number of service defects missed by ultrasonic inspections is historically about 3 times less for low-speed passenger trains, 10 times lower for moderate-speed passenger trains, and 100 times less for high-speed passenger trains.

12

Buckled Track

The City of New Orleans, made famous by Arlo Guthrie's 1972 hit song of the same name, departed its namesake at 1:55 p.m. on Tuesday, April 6, 2004, bound for Chicago. The Amtrak passenger train, consisting of one locomotive, one baggage car, and eight passenger cars, derailed in Mississippi about halfway between two bridges crossing the Big Black River at 6:33 p.m.

The engineer could see something wrong with the track ahead and immediately started slowing the train. Too late for a safe stop, the train reached the track buckle and started to derail. The engineer applied the emergency brakes and watched the right rail overturn in front of the skidding train.

The engineer recalled the accident: "I get up to it and I can see that the rail is kinked, and my engine goes through. I think the baggage car went through it too, and I started to feel everything moving and shaking. I knew something wasn't right. And my engine is already through it, and I'm not looking in the mirror, I'm looking ahead and I can see the right, or east rail rolling over, [so] I shot it" (i.e., he applied the emergency brakes).[1] The locomotive and first attached car successfully passed the track buckle and remained upright. The train broke apart behind the second car with the next six cars skidding to a stop on their side. The eighth car was tilted 45 degrees; the last car was upright.

The river bottom land, normally a swamp, was parched dry after weeks of drought. Minutes later, a dust cloud swirled from the accident site and into a neighborhood 1.5 miles (2.4 km) away. "You couldn't see anything, and you needed a mask."[2]

The locomotive's event recorder fills in the details. With the train traveling at 78 mph (125 km/h), the service brake was applied at 6:33:27 and the emergency brake 6 seconds later. During the next 13 seconds the train traveled 398 feet (121 m) before derailing to a stop. The train, derailing to the right, ripped out the right hand rail. The engineer saw the rail roll over in front of the locomotive. While the locomotive was skidding to a stop, a 40-foot (12-m) section of track came loose and speared a 68-year-old woman, the only fatal injury.

THE RESCUE

Fifteen minutes after the 911 call, the first emergency responders arrived on the scene about 6:52 p.m. Rescue workers had to locate the wreck, drive 1 mile (1.6 km) on a dirt road, and travel on foot the last half mile, lugging oxygen tanks, first aid kits, and other equipment. Later, all-terrain vehicles and railroad hi-rail trucks evacuated the injured.

The first responders found confusion, fear, bloody passengers on the embankments, and others trapped inside the overturned cars. "I just saw people jumping out of the train, screaming and bleeding," said the Kearny Park Volunteer Fire Department chief, one of the first people to arrive. "We had to smash the windows to get into the cars."[3]

Five minutes after arriving on the scene, the fire chief requested assistance from neighboring fire departments. Eventually, an estimated 200 rescue workers from 21 agencies (fire, police, and municipal and private ambulances) responded to the accident. Two days after the accident, Amtrak train service resumed, albeit at a reduced speed through the repaired accident site.

The track was owned by Canadian National Railroad. Amtrak was not responsible for track maintenance and had no direct responsibility for the accident. At that time the previous Amtrak fatality had occurred nearly 2 years and 10 billion passenger miles before.

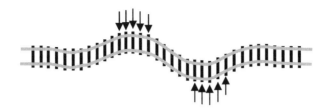

FIGURE 12.1. Common track buckle **S** pattern occurring in straight track. Arrows show the lateral forces required to prevent buckling.

TRACK BUCKLES

During hot weather rail thermally expands. If the rail is not properly constrained laterally, the expansion can buckle the track. Rail-buckling occurs in well-known patterns. The **S** pattern shown in Figure 12.1 is common on straight track, while a **C** pattern occurs on curves. Typical buckled deflections are 12 to 28 inches (38 to 71 cm) on straight track and 6 to 14 inches (15 to 35 cm) on curves.

Many cross ties near the City of New Orleans derailment site were split and would not hold their rail spikes. With thermal expansion, the rail bowed up and lifted the spikes out of the weakened ties. As the rails lifted off the ties, the track lost lateral support and buckled sideways, as shown in Figure 12.1. On the day of the accident, the temperature was 80°F (27°C). The rail temperature was estimated to be 110°F (43°C) on the bright sunny day.

Usually, the spikes hold the rail in place, preventing rail uplift. In that case the weak link to thermal expansion becomes the crushed rock ballast holding the cross ties in place. If the ballast lacks sufficient strength, the cross ties and track buckle sideways.

The next major work on this track was scheduled for early 2005. On an accelerated schedule after the accident (between May 2004 and July 2005) Canadian National replaced more than 52,000 cross ties and added crushed rock ballast to a 53-mile (85-km) section of track, including the derailment site.

RAIL JOINTS

Historically, rails did not buckle. Rails were installed in 39-foot (11.9-m) sections connected by bolted joints. The gap every 39 feet between the 2 rail ends accommodated thermal expansion.

The rail joint, initially a nineteenth-century strap to keep the rail ends aligned, developed into a structural element that attempts to make the connected rail ends as strong as continuous rail. The joint bars fail miserably. Nevertheless, the rail joint does provide strength that reduces rail stress and deflections (Figure 12.2).

As the wheels pass a rail joint, the rail ends deflect downward one at a time, creating a small step and mismatch in rail heights. The step causes premature wear from wheel impacts known as end batter. In time, the end batter requires repair by cutting off the rail ends or by rebuilding the rail height with welded metal. The rail joint also degrades as the bolts and supporting ballast loosen. Eventually, increased joint deflections cause freight cars to rock with more force on the joint, which accelerates wear. Freight car rocking can even derail the train (see "Harmonic Rock-Off" in Chapter 7).

In 1920, there were about 250,000 route miles (402,300 km) and 400,000 track miles (643,700 km) with more than 54 million rail joints every 39 feet (11.9 m) in the United States. The bolts would rattle loose and constantly need retightening. The joints were lubricated (and relubri-

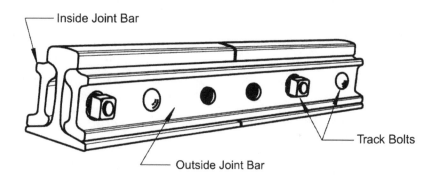

FIGURE 12.2. Joint bars connecting two rail ends. *From NTSB.*

FIGURE 12.3. Metal fatigue fracture originating at a bolt hole.

cated) to prevent them from corroding shut and accumulating thermal expansion that could buckle the track. The rubbing parts of the joint wore and eventually needed shimming. The rail joint was a constant maintenance headache.

But worst of all, the higher stresses at the bolt holes encouraged fatigue cracks (Figure 12.3). The joint bars also hid rail fatigue cracks, making them more difficult to find. Traditionally, 50% of all broken rail occurred at rail joints.

On November 5, 1967, a broken joint bar derailed a commuter train in London at 70 mph (113 km/h). A triangular piece of steel broke out of a rail joint and wedged under a wheel. The 12 cars were crowded with an estimated 1,000 passengers. Forty-nine passengers were killed and 27 seriously injured.

Metal fatigue can be explained as simply as bending an aluminum strip back and forth until it breaks. In fact, if one bends a section of aluminum mini-blind by hand, a fatigue crack can be seen to grow right

Fatigue at a Hole

A 0.1-inch (2.5-mm)-diameter hole was drilled in strips of aluminum mini-blind. The mini-blind was clamped between two 0.25-inch (6.3-mm)-diameter rods. The aluminum strip was bent up and down (±90 degrees) forcing the aluminum to bend to the radius of the rods. (Without the rods, a sharp crease formed during bending and significantly reduced fatigue life.)

A set of 3 tests without holes averaged 104 bend cycles (twice as many cycles compared to the strips with drilled holes). The fatigue cracks are seen to start at the hole where the stress is higher. A much bigger decrease in the fatigue life was expected for the specimens with holes. If the strips are bent, say ±10 degrees, it might take 1,000 cycles to failure (with holes) and 4,000 cycles (without holes) to demonstrate the point.

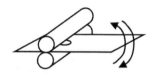

Fatigue at a hole, test setup.

before your eyes. A 1-inch (2.54-cm)-wide, 0.007-inch (0.18-mm)-thick aluminum strip repeatedly bent with (and without) a hole demonstrates how a hole reduces fatigue life.

RAIL JOINTS TODAY

Today most rail joints have been replaced with continuously welded rail. A conservative estimate, however, suggests that there are still more than 140,000 rail joints in North American main line service today. Bolted joints are used on temporary repairs, on curves with high wear, and on secondary track.

One such rail joint, a temporary track repair, failed from metal fatigue and derailed a 12,342-ton (11,200-MT), 7,138-foot (2,175-m)-long, 112-car Canadian Pacific train on January 18, 2002, in Minot, North Dakota, releasing a 5-mile (8-km)-long plume of anhydrous ammonia into the town of 35,000. Fortunately, only one person died in the zero-visibility, fog-like ammonia cloud. Everyone else safely stayed in their homes. Because of Minot and other similar derailments, there has been increased analysis of rail joints by the Federal Railroad Administration (FRA). In fact, since 2007, the FRA requires a written report for every single rail joint failure.

Unlike the high-speed inspection of continuous welded rail, rail joints must be inspected by hand. The FRA initiated research to improve rail joint inspection. A high-speed camera mounted on a vehicle can inspect rail joints while moving at 60 mph (97 km/h). The system locates each joint with lasers and then triggers a high-resolution camera. Advanced image-processing algorithms using pattern recognition software can detect even hairline fatigue cracks. The stated accuracy continues to rapidly improve.

THE ENTIRE RAILROAD INDUSTRY wanted to replace rail joints with continuously welded rail. Concerns about thermal expansion and buckling, however, held back welded rail for many years. Until proven with tests, no one thought a pile of loose rocks would prevent buckling. Testing did demonstrate the need for better lateral support. Consequently, closer tie spacing and a wider bed of crushed rock became standard.

Unsupported, long, skinny structures are in fact weaker in compression than in tension. A wooden yardstick attached to the ceiling can easily support a person's weight in tension. If that person leans just slightly on top of the same yardstick (sticking up from the floor), compression will quickly cause bowing and buckling.

The buckling load depends on the unsupported length. A 50-foot (15-m) length of 136 lbs/yd rail mounted like a flag pole will buckle when loaded with 77,500 lbs (344 kN) on top. An 80-foot (24-m) length of rail loaded the same way will buckle with only 30,300 lbs (135 kN). The weight required to fracture the same rail in tension is more than 2 million lbs (8,896 kN).

The yardstick and the rail will support much higher compressive loads if supported laterally. The support provided by the rail attached to cross ties embedded in crushed rock is expected to raise the buckling load to around 150,000–200,000 lbs (667–890 kN). Any disruptions to the crushed rock ballast will significantly lower the buckling resistance of the track.

CRUSHED ROCK BALLAST

Crushed rock ballast, underneath the cross ties, prevents the ties from sinking into the ground under the weight of the train. The ballast must be angular, with sharp edges that interlock. A pile of spherical

Capillary Action

Capillary action is as familiar as water wicking up against gravity into a paper towel. Capillary action can be explained as the liquid having a greater attraction for a surface it touches than with itself. Water on a surface can "bead up" or spread out and "wet" the surface. Attraction between water molecules is called cohesion. Attraction between liquid and a surface is called adhesion. Water beads up if cohesion forces exceed the adhesion forces. The cloud of electrons surrounding a molecule is not uniformly distributed, making every molecule a tiny magnet. The magnetism creates adhesion and cohesion forces.

marbles will not hold the track in place; in fact, the marbles will not even maintain a pile on a flat surface. Rocks with rounded edges are used on truck runaway ramps on steep hills. Rounded rocks do not interlock; the runaway truck sinks into the pile of rounded rocks and stops.

Large rocks are crushed to create the sharp edges. The sharp edges also bite into and grip the wooden ties. The ballast must be consolidated to function properly and to adequately interlock. To allow the ballast to consolidate it is common practice to restrict train speed after any disruption to the ballast. Ballast is disrupted when replacing cross ties or resetting the elevation of rail that has settled and sunk.

Many things can go wrong with the ballast. The crushed rock can become worn with repeated train movement and maintenance; the sharp edges become rounded and will not interlock.

The ballast must also provide good drainage. The voids between the rocks can fill up with particles, including coal dust falling off of coal cars, accumulated abrasion, or sand used to improve traction. Eventually, accumulated particles will prevent proper drainage and the ballast picks up moisture from capillary action. The result is a cross tie that pumps up and down in a mud hole every time a wheel passes.

Accumulated moisture can rot the ties as well as loosen and corrode the spikes. Also, if the underlying soil remains wet, the tracks will eventually shift. The problem of fouled ballast is so common that it is standard practice to periodically clean the ballast by digging out the crushed rock under the tracks and removing smaller particles with vibrating screens. Eventually, the ballast must be replaced, perhaps every 250 MGT or so.

In another recent technological advancement, the ballast is increasingly inspected by ground penetration radar. Pulses of radar are used to create a subsurface image that looks for problems with the ballast, subballast, and underlying dirt.

ANOTHER COMMON PROBLEM IS SETTLEMENT, which occurs when the rail and rock ballast shift downward. Settlement is especially a problem on curves where the outer rail is raised 4 to 7 inches (10 to 17.8 cm) to reduce inertial loading. The curving loads tend to force the two rails apart if the ties lack strength from rot.

When settlement occurs, it is common practice to lift the track to reset the track elevation. This maintenance work is known as "surfacing." When the rail and attached cross ties are lifted (perhaps 1 to 6 inches [2.5 to 15 cm]), an empty void occurs under the ties, as shown in Figure 12.4.

After the ties are lifted, a process known as "tamping" involves inserting two arms into the ballast on both sides of the cross tie. The arms are

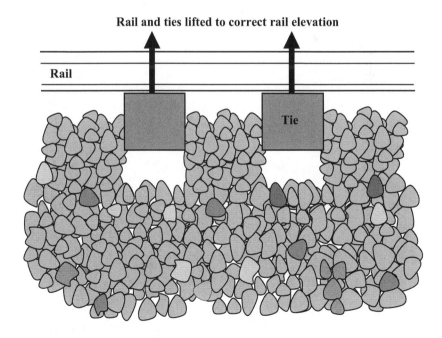

FIGURE 12.4. When rails and attached ties are lifted to reset rail elevation, a void is created under the ties. Tamping fills the void.

vibrated at 3,000 to 4,500 vibrations per minute to ease insertion into the crushed rock. (The vibrating crushed rock "liquefies," or at least acts more like a liquid.) The arms are squeezed together under the cross tie to fill the void and hold the tie in its new lifted position.

Tamping disrupts the consolidation of the ballast and the grip of the sharp edges with the wooden ties. Tamped ballast typically is 40% to 60% less effective in resisting track-buckling. The consolidation has to be reestablished mechanically with massive vibrators that simulate train traffic or by actual traffic. Tamped track usually has a "slow" order in place until a sufficient passage of traffic has occurred.

LATERAL STRENGTH

The lateral strength of one tie is measured with the Single Tie Pull Test. One tie is disconnected from the track and hydraulically pulled. It takes 3,000 lbs (13.3 kN) to pull a wooden tie in well-compacted, crushed rock. Track maintenance can lower the tie's pull strength to 1,000 lbs (4.4 kN).

Many buckled track derailments are, in fact, associated with weak-ened ballast from recent track maintenance. Ballast can also shift from earth-moving equipment repeatedly driving over it.

When the ballast is worn out, it is common practice to dump new ballast on top of the old ballast, lift the track above the new ballast, and tamp under the ties. The net effect is that the pile of ballast accumulates and the tracks are slowly lifted higher and higher. As the pile of rocks gets higher, there may not be enough rock on the outer edges of the cross ties to hold them in place, as shown in Figure 12.5. The crushed rock ballast should be 12 to 18 inches (30 to 46 cm) deep on the outside of a curve, holding the cross ties in place. This exact problem occurred with Amtrak Train P052-18 near Crescent City, Florida, in 2002.

AMTRAK TRAIN P052-18

The Amtrak train left Sanford, Florida, on April 18, 2002, bound for Virginia. The train consisted of 2 locomotives, 16 Amtrak passenger cars, and 24 Amtrak "autorack" cars loaded with the passengers' cars.

Entering a curve at 57 mph (92 km/h), the engineer saw a 10-inch

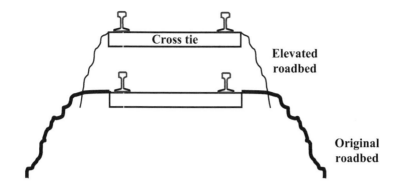

FIGURE 12.5. Elevated roadbed has insufficient ballast on the sides of the cross ties.

(25-cm) track buckle immediately ahead. While reaching for the brake handle, the engineer was violently jarred when the locomotive passed the buckle. An Amtrak attendant said he felt the train "starting to zigzag" and then it toppled. "The tracks had come loose, like thread. They were turned all different ways, and the wood was shredded."[4]

A passenger recalled that "Suddenly the train bucked as if someone stepped on the brakes." He asked the bartender if this was normal. The response was interrupted by "the sound of cracking wood." The bartender picks up the story: "The train lurched first to the left, then to the right, and up in the air. I was thrown from one side of the car to the other. I was knocked out."[5]

The first locomotive came to a stop about 700 feet (213 m) after the engineer applied the emergency brakes. (According to the event recorder, the service brake was applied at 5:07:57 p.m., the emergency brake 4 seconds later.) The 3rd through 23rd cars derailed. The first 8 derailed cars were lying on their sides parallel to the track. The next 7 cars were "zigzagged" accordion-fashion perpendicular to the track. Cars 17 through 23 were derailed upright. Violent impacts occurred when the cars in the back continued to slam into the cars in front, which were now sideways off the track.

The overturned cars were dragged on their sides for some distance. Four passengers were killed after falling out of broken windows and 35 were seriously injured. Twelve people were evacuated by helicopter to hospitals in Jacksonville, Daytona Beach, and Gainesville.

THERMAL EXPANSION

When heated, the steel rail expands; if cooled, it contracts. On an atomic scale all atoms above absolute zero (−459°F [−284°C]) are vibrating with thermal energy. More heat energy increases the vibrations and the distance between atoms. When the steel is cooled, the vibrations and the atomic spacing decrease.

One inch (2.54 cm) of steel heated 1°F will expand 0.0000065 inch (0.000165 mm). This value is called the coefficient of thermal expansion for steel. The total expansion increases with the length and the change in temperature. For example, 100 inches (2.54 m) of steel heated 50°F will expand 100 × 50 × 0.0000065 = 0.0325 inch (0.089 cm). The process works in reverse. If the same length of steel is cooled 50°F, the steel will contract 0.0325 inch. In equation form:

$$\text{change in length (inches)} = \text{rail length (inches)} \times$$
$$\text{temperature change (°F)} \times 0.0000065 \, \frac{\text{inches}}{\text{inch °F}}$$

RAIL ANCHORS

Rail anchored on both ends is prevented from expanding or contracting during temperature changes. Consider a 100-inch (254-cm) span of rail removed from its wooden ties and anchored at room temperature (70°F [21°C]) on both ends to extremely rigid walls that prevent all thermal movements (Figure 12.6).

If the rail remains at 70°F (21°C), there are zero stresses in the rail. If the rail heats up by 10°F (5.56°C) (to 80°F [27°C]), the rail tries to ther-

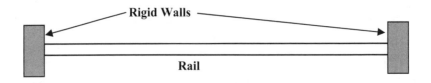

FIGURE 12.6. Thermal movement is prevented by attaching both ends of the rail to rigid walls.

mally expand. The rigid walls prevent the expansion by exerting a compressive force that exactly cancels out the attempted thermal growth. If the rail heats up by 20°F (11°C), the thermal expansion is twice as much and the force from the walls to prevent the thermal expansion also doubles.

If the rail instead cools down relative to the installation temperature, the rail tries to contract. The rigid walls force the rail to remain at the same length by exerting a tensile force that exactly cancels out the attempted contraction.

Anchored rail heated creates compressive forces that can buckle the rail. Anchored rail cooled will create tensile forces that can fracture the rail. The heat-up or cool-down is relative to the installation temperature. In the winter rail tends to pull apart and fracture. In the summer rail tends to compress and buckle.

Buckling from thermal expansion is easily demonstrated with a 6-inch (15.2-cm) strip of aluminum bar 0.5 inch wide and 0.125 inch thick (1.27 × 0.31 cm).[6] If placed between rigid supports (a C clamp will do) and heated with a torch, the aluminum bar will quickly buckle.

RAIL IS NOT ANCHORED by extremely rigid walls on each end. The rail sits on tie plates that are attached to the wooden ties (sometimes concrete ties are used) with spikes.[7] Thermal movements in the rail are prevented by using rail anchors. The rail, tie plates, spikes, and rail anchors are shown in Figures 12.7 and 12.8.

To properly prevent longitudinal rail movement, the anchors must be snug against the cross ties. Inspection after the City of New Orleans derailment found about 50% of the rail anchors were either missing or not snug.

Thermal loads from the rails can be more than 200,000 lbs (889 kN). The small anchors cannot support this force. One anchor might resist 500 to 1,200 lbs (2.2 to 5.3 kN). The large thermal loads must be shared across many rail anchors. To create an effective anchor point that resists all thermal contraction and expansion at the ends of long spans of rail requires using rail anchors at every cross tie for 195 feet (59 m). Rail is anchored at bridges, turnouts, crossings, track switches, and the like to protect these structures from high thermal loads.

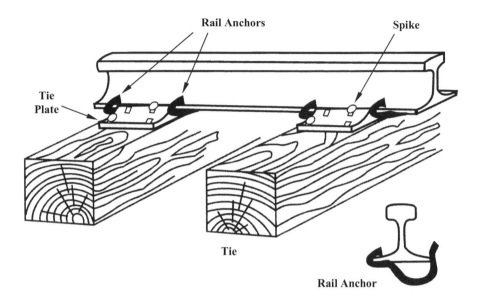

Rail Anchors

Spike

Tie Plate

Tie

Rail Anchor

FIGURE 12.7. Rail anchor clips, spikes, tie plate, tie, and rail. *Adapted from NTSB/ RAR-03/02.*

FIGURE 12.8. Rail, tie plate, spike, and anchor clips.

In between anchor points rail anchors are installed on every other tie to limit thermal contraction if the rail breaks and to limit rail creep that occurs when the train accelerates or brakes.

RAIL FORCES

Consider a 1,500-foot (457-m) span of rail held in place with multiple rail anchors on each end. The rail anchors collectively resist any thermal expansion or contraction. The rail's attempted contraction can be calculated by the coefficient of thermal expansion multiplied by the length of the rail and the change in temperature relative to the installation temperature. If the 1,500-foot length of rail is cooled to 50°F (10°C) lower than its installation temperature, it will try to contract 5.85 inches (14.8 cm).

The anchors automatically create a tensile force that elongates the 1,500-foot rail by 5.85 inches to exactly cancel the attempted thermal contraction. The thermal contraction force in the rail equals the force required to stretch the same length of rail by 5.85 inches.

If the 1,500 foot × 12 = 18,000-inch (457-m) span of 136 lbs/yd rail (cross section 13.35 in^2 [86 cm^2]) is supporting a 130,160-lb (59,000-kg) weight (Figure 12.9), the rail will stretch 5.85 inches (14.8 cm).[8] Double the load and the stretch doubles. The ratio of load divided by stretch is a constant for this 1,500-foot length of 136 lbs/yd rail.

Engineers prefer to think in terms of *stress* and *strain*, both engineering terms. The stress in this rail, or force per unit area, equals 130,160 lbs/13.35 in^2 = 9,750 lbs per square inch, or 9,750 psi (67.2 MPa). The strain in this rail is the elongation divided by the initial length or (5.85 inches)/(18,000 inches) = 0.000325. The ratio of stress divided by strain (called the modulus of elasticity) is a material constant for all steel rails of any length and cross sectional area loaded with any load.

The stress divided by the strain is a material constant for all steels because it describes the stretchiness of the iron atomic bonds (steel mostly consists of the element iron) and equals about 30 million psi (206,800 MPa). Knowing this constant makes calculation of the rail thermal force relatively easy.[9] A plot of the tensile force in a 136 lbs/yd rail versus temperature decrease is shown in Figure 12.10.

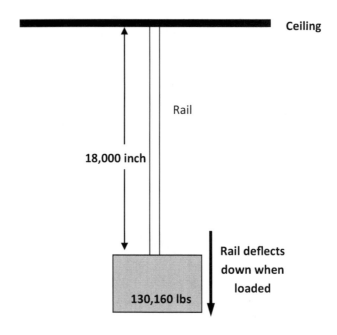

FIGURE 12.9. Hanging 136 lbs/yd rail loaded with a 130,160-lb weight deflects down 5.85 inches.

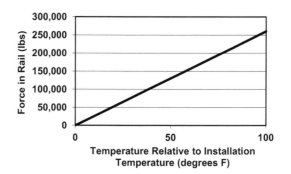

FIGURE 12.10. Force in the rail versus cooling relative to installation temperature.

RAIL INSTALLATION TEMPERATURE

The rail installation temperature, also called the rail neutral temperature, is the temperature at which zero thermal stresses exist in the rail.[10] We will also use the term the rail's *stress free temperature*.

Since tensile stresses in the winter can fracture the rail and compressive stresses in the summer can buckle the rail, one might think the ideal rail installation temperature is the average temperature. Because summertime buckling is a greater concern, the rail's stress free temperature is set higher than the yearlong average.

If the rail contracts by 60°F (15°C) during the winter, a tensile stress of about 11,700 psi (81 MPa) results, a very small value compared to the tensile strength of the rail steel—about 160,000 psi (1,100 MPa). The rail will not fracture during the winter without a slowly growing fatigue crack that will hopefully be discovered during normal inspection. Surprisingly, only about 1.4% of undetected broken rails actually derail a train.

Buckled track, on the other hand, has a much higher probability of occurring suddenly and derailing the next train. Also, radiant heating from the sun can heat the rail 30 to 40°F (−1 to 4°C) hotter than the ambient temperature on a sunny day. For these reasons the installation temperature is much higher than the average temperature and designed to mostly protect against rail-buckling. In North America, the stress free temperature varies from 90 to 105°F (32 to 40°C), depending on local temperatures. The City of New Orleans derailment site in Mississippi had a target stress free temperature of 105°F (40.5 °C).

THE STRESS FREE TEMPERATURE DEGRADES

The rule of thumb is that rail has to heat up 60 to 80°F (15 to 27°C) before buckling occurs, an unlikely proposition for rail installed at 105°F (40°C). When the rail's stress free temperature degrades from 105 to 60°F (40 to 15°C) or less, buckling can occur.

On very cold days, the rail tries to contract. Rail installed with a stress free temperature of 105°F (40°C) will have about 260,000 lbs (1,156 kN) of tensile force when cooled to 5°F (−15°C). If the anchors provide no resistance to thermal contraction, the rail will contract and act like it was installed at 5°F (−15°C). This rail will buckle when it heats up to 65 to 85°F (18 to 29°C). If the rail anchors are missing or not snug and cannot provide 260,000 lbs of resistance, the rail will contract and act like it was installed at a temperature lower than 105°F (40°C).

Curves are especially troublesome. During winter contraction, the rail on curves needs sufficient lateral support. If the ballast does not hold the

FIGURE 12.11. Curved track pulls in to dashed lines shown during contraction in the winter. This lowers the rail's stress free temperature.

cross ties in place, the track pulls in on curves, as shown in Figure 12.11. This pulled-in curve will now buckle at a lower temperature.

Track repair is the most common problem causing the stress free temperature to degrade. Whenever the track is lifted at temperatures lower than the stress free temperature (to reset rail elevation), the bond between the crushed rock and cross ties is broken. The rail is then free to contract and bow inward slightly on a curve, thus releasing the rail's tensile force. The derailment in Crescent City, Florida, occurred on curved track that had been lifted in March. The derailment occurred in April during the first hot spell of the season.

ANOTHER COMMON MAINTENANCE PROBLEM is replacement of damaged rail at temperatures below the stress free temperature, one of several problems leading to Amtrak's City of New Orleans derailment described at the beginning of this chapter. A section of damaged track 12 feet, 11.5 inches (3.95 m) long was cut out at 60°F (15°C) in January, two months before the accident. At 60°F (15°C) the track, trying to contract relative to the stress free temperature of 105°F (40°C), was in tension. When the track was cut, tensile forces were released and the 12-foot, 11.5-inch (3.95-m) gap grew to 13 feet, 2 inches (4.01 m) (Figure 12.12). The rail was now acting like it had been installed at a lower temperature. The stress free temperature for the remaining track in a situation like this is difficult to determine and depends on the number of rail anchors, how snugly they fit against the cross ties, the condition of the ballast, and how much tensile force is retained in the rail after it is cut. (Recall that the forces required to rigidly anchor the rail against thermal expansion are spread over numerous rail anchors.)

For the new 13-foot, 2-inch section of replacement track, the stress free temperature was the temperature it was installed at, 60°F. The replacement track thermally expanded on the next hot day and buckled.

One correct procedure for the repair of this track on cold days would

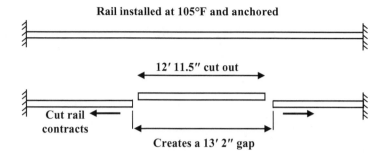

FIGURE 12.12. Rail cut at 60°F (16°C) contracted as shown.

have been to release the rail anchors for 200 feet (61 m) on both sides of the cutout and heat the rail to shrink the 13-foot, 2-inch gap back to its original length of 12 feet, 11.5 inches.

Because the dispatcher gave the repair crew at the City of New Orleans derailment site only 1 hour to make the repair there was not enough time to release and reinstall 400 feet (122 m) of anchors. The quickest and easiest repair was to bolt in a 13-foot, 2-inch section of replacement rail with rail joints. This quick repair made the track susceptible to buckling in the summer. Everyone understood the problem. The solution would have been to take out the repair and make a proper repair (with welded rail joints) before the summer heat-up. The paperwork got misplaced, and the rail was not properly adjusted before the seasonal heat-up.

TRAIN FORCES ON THE TRACK

Although a rail can buckle by itself, it is more common for buckling to be triggered by the lateral forces from a passing train. Buckling is also very sensitive to track misalignment. Misalignments are particularly bothersome on curves, where, in fact, most buckles occur. Recall that the wheels always exert a lateral force on the outer rail of curves (see Figure 7.6 in Chapter 7). The lateral wheel force increases with increased track misalignment. The permitted misalignments (deviation from a 62-foot [18.9-m]-long string pulled tight) for the different class tracks are shown in Table 12.1. Recall that higher class track corresponds to higher allowable speeds. French high-speed trains (up to 200 mph or 322 km/h) limit track misalignments to 0.157 inch (4 mm).

TABLE 12.1
Permitted Misalignment

Class of Track	Permitted Misalignment (inches)
1	2″
2, 3	1.75″
4, 5	1.5″
6 and higher	0.5″

Buckling on curves tends to be incremental, as unstable track shifts a tiny amount with every passing train. Eventually, the lateral movement becomes excessive; the compressive rail force takes over and suddenly buckles the track.

Often the front of the train will incrementally shift the track until a derailment occurs in the back of the train. A Canadian study of 18 track buckle derailments found that 15 occurred in the second half of the train and 13 in the last 15 cars of the train.

Just two hours before the City of New Orleans derailed in 2004, a freight train had passed over the site at 4:30 p.m., the hottest time of the day. The track probably buckled toward the back of that train. The faster, lighter Amtrak passenger train could more easily bounce off the tracks than the earlier, slower 17,000-ton (15,422-MT) freight train.

With straight track the long runs of rail can accumulate more compressive energy, just as a longer coiled spring contains more energy than a shorter one. For this reason straight track buckling can be very rapid and spontaneous.

BUCKLING TESTS

A 7.5-degree curve (562-foot radius) with 136 lbs/yd rail was tested for buckling. The cross ties initially had a lateral strength of 2,500 lbs (11.1 kN) per tie. Setting the lateral misalignment at 1.5 inches (3.8 cm) (the maximum for Class 4 rail) disrupted the ballast and reduced its lateral strength to about 1,350 lbs (6 kN) per tie. The rail was electrically heated until it explosively buckled after a 66°F (19°C) heat-up. The buckle shifted the track laterally 13.5 inches (34 cm) over a 40-foot (12-m) span. Overnight cooling reduced the misalignment to just 2.5 inches (6.3 cm).

After repairing the buckled track and strengthening the ballast up to

1,780 lbs (7.9 kN) per tie, dynamic buckling tests continued with heated rail and the passage of an actual test train with 1 locomotive and 24 loaded hopper cars. The track was set with an initial misalignment of 0.75 inch (1.9 cm) on the curve. The rail was heated 32°F (0°C) above its installation temperature. The train's passage at 34 mph (54 km/h) shifted the rail an additional 0.25 inch (0.6 cm) for a total misalignment of 1.0 inch (2.5 cm).

Before the train passed a second time, the rail was heated 48°F (9°C) above the installation temperature. The misalignment progressively grew under the passing train from 1.0 inch to 1.5 inches (2.5 to 3.8 cm). On the final train run, the rail was heated to 62°F (17°C) above installation. The rail progressively buckled under the passing train until a maximum displacement of 10 inches (25.4 cm) derailed 4 cars. During the final test, 2 new buckles (6 and 8 inches [15 and 20 cm]) suddenly appeared. Two additional cars derailed at the 8-inch buckle.

Surprisingly, buckled track does not always derail trains. In June 2006, an Amtrak train in Louisiana braked from 67 mph (108 km/h) and safely traversed a 25-inch (63-cm) track buckle.

BUCKLING ANALYSIS

Buckling of structures is very important in many fields of engineering; the columns of a building, for example, must not buckle. All of the variables associated with track buckling can be analyzed with computer models developed by the FRA and the railroad industry. Computer simulation is now widely used to plan inspection and track repairs that reduce the likelihood of buckling.

Buckling is usually associated with:

1. Reduced lateral resistance of the cross ties
2. Lateral misalignments in the track
3. Lowered stress free temperature of the track
4. The first hot and sunny day of the season

Railroad companies often reduce speed and increase track inspections on excessively hot days. Reduced speed lowers train forces on the track and lessens potential damage from a derailment.

Buckles also tend to happen at the bottom of hills, where braking occurs in the same location (braking pushes and compresses the track ahead), and next to immovable objects (i.e., bridges) that are securely anchored.

Presented in Figure 12.13 is an FRA-developed plot comparing buckling on a 7.5-degree curve and on straight track for 136 lbs/yd rail. The stress free temperature in Figure 12.13 has degraded to 70°F (21°C) and the track is misaligned by 1.5 inches (3.8 cm) (the limit for Class 4 track used by freight trains up to 60 mph [97 km/h]).

Buckling is predicted when operating below the 2 lines. Figure 12.13 predicts a buckle on the curve when the 70°F (21°C) rail heats up to about 133°F (56°C) and the lateral resistance of the ties is 2,200 lbs (9.8 kN) or less (see star on plot). Straight track is shown to be considerably more stable than curved track.

The computer model is very good at sorting out the variables of misalignment, rail expansion, lateral constraint, and train speed. The problem is that many of these variables are difficult to accurately know.

Misalignment is measured with special geometry cars, but the track is always shifting and it's difficult to have current information. The track, however, is expected to be in compliance with the maximum permitted misalignment for the given track class.

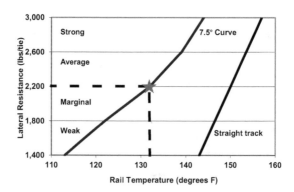

FIGURE 12.13. Relationship between rail heat-up temperature and required lateral resistance to prevent buckling (based on computer simulation) on a 7.5-degree curve and straight track for 136-lb track with a 1.5-inch lateral track misalignment and a stress free temperature of 70°F (21°C). *Adapted from DOT/FRA/ORD-93/26.*

The lateral strength of the cross ties is very difficult to identify without hydraulically pulling individual ties. The trained eye can spot potential problems with a visual inspection. Rail temperature is related to ambient conditions, but heat-up from the sun is more problematic.

The biggest unknown is determining the current stress free temperature in the track after repairs, shifting, and winter pull-ins. Numerous systems are being developed, tested, or used commercially. One device measures the thermal stress in the rail by disconnecting the spikes and hydraulically lifting the rail. Just as a guitar string is harder to pluck as it gets tighter, so is the rail. By measuring the force required to lift the rail, the thermal forces can be accurately determined. This system works very well but is highly labor intensive and cumbersome. Also, if the rail is in compression when lifted off the rails, it will bow and become misaligned when dropped back on the cross ties.

Some railroads have tested strain sensors and a thermocouple mounted at the time of a repair (or brand-new installation) to measure changes in neutral temperature. The disadvantage is that it can only measure changes relative to a known installed temperature. It will not measure a degraded stress free temperature on already installed rail. Research continues on advanced technology to solve this problem.

When track is cut to make a repair, a gap appears if the rail is cut below its stress free temperature. Computer programs now exist that can calculate the track's existing stress free temperature based on the width of the gap, ambient temperature, and assumed anchor forces. The computer program advises how to reset the track's stress free temperature to the correct value. The computer's output includes how many anchors to disconnect, how much rail to cut out, how to heat or pull the rail to close the gap, and the anchoring pattern after repair.

Safety in the Modern Era

RAILROADS AND THE ELECTRONIC AGE

Amazingly, the basic idea of a steel wheel rolling on a steel rail, which has remained in place for nearly 200 years, has hardly been affected by contemporary electronics. Modern electronics occur on the locomotive and all around the tracks but not on freight cars. Other than one radio-controlled sensor in the back of the train to activate emergency brakes and increasingly sophisticated computer controls on the locomotives up front, no other electronics exist in between.[1]

Excluding the signals used to safely separate trains, most electronics located off the train are intended to find unsafe equipment. In the past, equipment was simply operated until it failed. Within the limitations of human vision, inspection of track and equipment has historically only found large defects just short of failure. The modern approach is proactive and designed to find defects before failure.

Trackside sensors look for overheated bearings and wheels, out-of-round wheels, and dragging equipment. Machine vision is increasingly used to monitor wheel wear, brake shoe wear, and freight car damage. Sensors to detect shifted loads and excess height (or width) are commonly placed before tunnels and bridges. High water, rock slide, earth-quake, avalanche, and high wind detectors are located as needed.

Today rail is inspected by high-speed cars that look for internal de-fects. Other cars use sophisticated optical and electronic sensors to in-spect for shifted rails. Computer models predict a train's response to track deviations and pinpoint needed track repairs. Ground-penetrating radar detects drainage problems that often predict track movements. And other inspection cars load the track sideways to locate weak cross ties.

RAILROADS AND THE COMPUTER AGE

Space-age computer simulation is increasingly used to improve component design. Computer models are used to increase reliability of couplings, wheels, bearings, brakes, axles, and track. Train dynamic mod-eling identifies unsafe combinations of speed, weight, length, and routes. Computer simulation to improve survivability during collisions remains an ongoing area of research and development.

The modern approach is to do a "hazard analysis." Teams of experts, including train operators and repair crews, try to anticipate what can go wrong and proactively intervene. Follow-up requires noting what ac-tually did go wrong and making adjustments as needed. For example, trains carrying hazardous materials down a dangerous hill and into a neighborhood would be off the charts with potential hazards. In that case many additional operating and maintenance safeguards would be put in place.

WHILE NOT THE FOCUS OF THIS BOOK, "human factors engineering" is a new discipline of increasing importance. The human-machine inter-face is studied for sources of error. Layout and visibility of displays and controls, presentation of warnings, information overload, and boredom are all scientifically studied, as are the effects of noise, ventilation, tem-perature, and vibration on human fatigue.

And then there is Positive Train Control, which promises to finally address human errors with computer intervention. Unfortunately, Posi-tive Train Control will not prevent equipment failures, the major source of derailments.

NOTES

Chapter 1. The Railroad Industry (as Seen through Accidents)

1. In 2002, railroads had about seven times fewer fatalities per trillion ton miles than did trucks.

2. Locally manufactured goods are shipped twice, once for the raw material and again for the finished product. Imported goods are shipped once.

3. Ellig, "Railroad Deregulation and Consumer Welfare," 144.

4. A Class I railroad in 2009 had revenue exceeding $379 million. Class I railroads typically carry about 95% of all freight traffic and are regulated differently. Because of mergers and bankruptcies, the number of Class I railroads in the United States has declined to seven.

5. Because of traffic congestion, commuter trains have recently flourished, growing 140% between 1970 and 2007. Over that same period, Amtrak intercity passenger traffic remained stagnant.

6. For example, two crowded commuter trains in India collided head on, exploded, and killed 286 in 1999. Both trains were traveling about 56 mph (90 km/h) and were carrying more than 1,000 passengers.

7. In 1985, in Ethiopia a 4-coach train derailed off a bridge into a steep ravine, killing 428. In 2005, a Japanese commuter train derailed speeding on a curve, slammed into an apartment building, and killed 106.

8. FRA, The Railroad Safety Program—100 Years of Safer Railroads, 21.

9. The FBI takes over the investigation if any criminal evidence is found.

Chapter 2. How Trains Crash, Then and Now

1. The youth receiving the text message was located and interviewed by the crash investigators. The engineer was making plans to meet the youth and let him drive the train.

2. The radius of curvature of the track was only 955 feet (291 m), a sharp curve for most railroad applications.

3. "A Quiet Trip Home," *Los Angeles Times*, September 13, 2008, A1.

4. "Carnage in Chatsworth," *Los Angeles Times*, September 13, 2006, A1.

5. "A Quiet Trip Home," *Los Angeles Times*, September 13, 2008, A1.

6. Besides the Metrolink engineer, one passenger died in the second passenger car. The location of the 25th fatality was undetermined.

7. NTSB Survival Factors Factual Report, Appendix F.

8. NTSB Survival Factors Factual Report, Appendix H.

9. Many locomotives have videos to protect the railroad during road-crossing accident litigation. The video is expected to record a car driving around the crossing gates and into the freight train. Because the train cannot stop, it always has the right of way—even with a fire truck on its way to a fire.

10. More than 1,000 firemen, police, and sheriffs responded to the accident.

11. There were reports of bystanders wandering in to impersonate injured passengers.

12. Backboards are used by paramedics to immobilize the spine of a trauma victim with a potential spinal injury.

13. "A Massive Response," *Los Angeles Times*, September 14, 2008, A1.

14. NTSB Survival Factors Factual Report, Appendix L.

15. NTSB Survival Factors Factual Report, Appendix L.

16. Fatalities for historical train wrecks are difficult to document and often in dispute, especially in foreign countries. Severe fires or falling into water often make accurate statistics problematic.

17. A 3.3% grade drops 3.3 feet for every 100 feet of track.

18. Major commercial planes are struck by lightning on average once a year without any danger to the passengers. The negative electrons repel each other and move to the outside of the fuselage shell. The inside surface remains relatively electron and charge free.

19. More than 200 wooden cars were still in service in Chicago in 1956.

20. A trash bag on fire (maximum release of almost 700,000 BTU/h) filled with 6 lbs of crumpled newspaper continued to spread a fire when placed on a seat or in a corner.

21. Thermal efficiency is the ratio of a machine's useful output energy divided by the input combustion energy. If there was no mechanical friction and somehow no heat was lost out the exhaust, the thermal efficiency would be 100%.

Chapter 3. Moderate-Speed Passenger Train Collisions

1. Brakes are explained in Chapters 9 and 10. Briefly, both electro-pneumatic and conventional airbrakes use air pressure to press the brake shoes against the wheels. Conventional brakes use an air pressure control signal. Electro-pneumatic brakes use an electrical signal.

2. During a test a Highliner car traveling at 16 mph (25.75 km/h) collided with a standing Highliner car. The cars did not override, but the collision did permanently bend the car body and underframe. Only minor injuries would have been expected.

3. The train's kinetic energy ($0.5 \times mass \times velocity^2$) at 30 mph (48 km/h) equals 19 million foot lbs (25,760 kJ).

4. Acceleration is discussed in more detail in Chapters 7 and 9.

5. A 3-inch (76.2-mm)-thick steel plate 25 feet wide by 18 feet high (7.62 m by 5.5 m) is backed by 2 feet (0.6 m) of reinforced concrete. The concrete wall is reinforced by 3 concrete vertical walls 2 feet (0.6 m) thick by 36 feet (11 m) deep and 18 feet (5.5 m) high. In between the 3 walls are 2 additional concrete walls 18 inches (0.45 m) thick, 20 feet (6.1 m) deep, and 8 feet (2.44 m) high. More than 1,000 tons (100,000 kg) of compacted soil were placed between the concrete walls.

6. NTSB, Collision Between Two Washington Metropolitan Area Transit Trains on November 3, 2004, 48.

7. Only in the modern computer era has it been possible to use electronic sensors to record thousands of data points per second. Even then, the actual crash motion is difficult to confirm. Components may crush many feet and then rebound a foot or two. High-speed cameras can only record a 2D projection of the 3D motion, making it difficult to accurately determine the actual 3D movements.

8. Amtrak wanted to test a high-speed TGV French train set. The French National Railways (SNCF) turned them down, stating that their prized train was a system and that operating it on the Northeast Corridor would be like "asking an opera diva to sing in a nightclub." The difficulties of operating trains on the Northeast Corridor are discussed in Chapter 5.

Chapter 4. Freight Train Collisions

1. There are no known statistics for engineers jumping off of trains. There are many documented examples available, depending on whether the FRA or NTSB happened to study that collision and write a report. The Collision Avoidance Working Group studied 65 collisions for the FRA and concluded that no recommendation could be made about staying or jumping.

Chapter 5. Avoiding Collisions

1. One locomotive engineer told me that he will occasionally get a single track warrant (the "golden ticket") for his entire 187-mile (301-km) route in rural Minnesota.

2. In 1909, there were about 650 freight trains (with 30,000 freight cars) and 1,000 passenger trains (with 300,000 passengers) operating between Buffalo and New York City each day. The De Witt Yard, the busiest in the world at that time, switched 8,000 to 10,000 freight cars per day.

3. Although an extremely rare event, track circuits have failed. One such failure occurred on the Washington Metro on June 22, 2009, resulting in a collision that killed nine. The electrified Metro cannot use the simple DC track circuit described above because DC power travels in the rails. Instead, an alternating current (AC) track circuit signal is superimposed on top of the DC power signal. Unfortunately, the track circuit's AC power transmitter created a spurious signal that confused the track circuit.

4. A 13- to 25-watt bulb, seen for nearly a mile (1.6 km) during the day (when focused into a narrow beam with special lenses), came into use around 1915. Light emitting diodes (LEDs) are now beginning to replace filament bulbs. The sun directly behind a signal light remains a problem.

5. The track warrant territory described earlier is an example of dark territory.

6. Private email from Mike Brotzman on April 14, 2009.

7. "Crash Comes at Crossing," *New York Times*, February 28, 1921, 1.

8. "Engine Crew Held for Porter Wreck," *New York Times*, March 5, 1921, 15.

9. Today the busiest train stations are in Tokyo. Penn Station (in New York City) is the busiest in North America. Grand Central remains the largest train station in the world based on number of platforms.

10. *National Geographic*, April 1923, 401.

11. Congress ruled that separating the road from the tracks with an overpass would cause excessive damage to existing wetlands at five locations. In 2005, three people died at a road crossing in Connecticut when an Acela train hit their car; this is the only Acela road-crossing accident to date.

12. A 2007 survey by the FRA reported 51% of trains (not tracks) controlled by CTC and an additional 14.5% operate with an automatic block system.

13. A track with 1,000 trains per year and 10,000 tons per train represents 10 MGT per year. The computer model considered dispatcher effectiveness, a more nebulous human quantity.

14. Passenger rail has more than 60% of the market for all travelers between New York City and Washington, D.C.

15. The wire on the windings inside the motors must be insulated to prevent short-circuits. With increased voltage, the required insulation becomes too heavy. Insulation weight places a practical limit on maximum motor voltage.

16. A conventional steam locomotive pulling 20 cars at 60 mph (97 km/h) requires 885 horsepower (660 watts) and 5,540 lbs (24.6 N) of force just to overcome air friction. The same train at 100 mph (161 km/h) requires 4,100 horsepower (3,057 watts) and 15,380 lbs (68.4 N). Work and horsepower are explained in Chapter 9.

17. Human fatigue and "human factors" (not a focus in this book) are major topics of ongoing research.

18. NTSB, Collision between Union Pacific Freight Trains MKSNP-01 and ZSEME-29 near Delia, Kansas, July 2, 1997. Washington, D.C.

Chapter 6. Positive Train Control

1. Test data from the U.S. Air Force measured an average response time to a caution light between 1 and 4.5 seconds, with an average of 2.4 seconds.

2. The manufacturer states that the Amtrak AEM-7 locomotives will stop from 120 mph (193 km/h) with 1.75 mph/sec (0.53 m/s²) deceleration with normal brakes and 2.25 mph/sec (0.68 m/s²) during emergency braking. An Amtrak test train (1980) at 120 mph (193 km/h) stopped within 6,900 feet (2,103 m) on dry rail pulling 6 cars during normal braking. This test did not consider a reaction time.

3. A train needs permission from the dispatcher to enter a main track. The dispatcher also defines the train's endpoint and where the train may need to stop and wait for other trains to pass. These orders are known as the train's authority limits.

4. ETMS (Electronic Train Management System) by BSNF, CBMT (Communications Based Train Management) by CSXT, ITCS (Incremental Train Control System)

by Amtrak, OTC (Optimized Train Control) by Norfolk Southern, and Vital Train Management System by Union Pacific are examples of specific PTC systems.

5. The exact quote: the cost of installing PTC "may be so high as to not be undertaken and therefore result in the elimination of Amtrak service" (*Wall Street Journal*, October 26, 2009, A5). Today Amtrak is aggressively pursuing PTC along with the rest of the railroad industry.

6. Many people may be uncomfortable with assigning a price to a human life. The author contends that an indirect cost-benefit analysis occurs every time one enters an automobile. More than 30,000 people annually are killed in the United States in automobile accidents. This author believes that this number could be reduced by 90% if we all drove $2 million safety-enhanced cars. Eventually, everything comes down to cost.

7. The proposed new PTC technology does not actually keep track of the back end of a train, which can be 1 to 2 miles (0.6 to 1.2 km) from the front. Without actually knowing where the back of the train is, the new PTC will not be able to prevent a collision if a train is allowed to enter an occupied block at reduced speeds.

8. This includes 94,313 miles (151,800 km) by Class I freight railroads (the largest railroads that carry more than 90% of the freight in the United States), 16,930 miles (27,246 km) by regional freight railroads, 28,891 miles (46,500 km) by local freight railroads, and 21,708 miles (34,900 km) by Amtrak (for the year 2007). The track miles are down from historic highs of about 429,000 track miles (690,400 km) (250,000 route miles [402,000 km]) in 1930.

9. NTSB accident report.

10. Track circuits also detect broken rails, a function not currently solved by radio-controlled, PTC-operated trains. Currently, freight trains are not allowed to operate above 49 mph (79 km/h) and passenger trains above 59 mph (95 km/h) without track circuits in place.

11. The rubber seals shrink in cold weather. Many companies operate shorter trains in colder weather to prevent excessive leakage and erratic braking. One such example is 150 cars at 15°F (−9°C) versus 80 at −16°F (−27°C).

12. One failure out of a million is 99.9999% reliability. If a radio transmission fails 1% of the time from interference, it is 99% reliable. The probability of 3 radio transmissions in a row all failing is $1/100 \times 1/100 \times 1/100 = 1/1,000,000$, or one radio failure per million tries, a reliability of 99.9999%.

13. Historically, a gyroscope was a rotating disk mounted so that it could take any orientation. When a locomotive tilted on a curve, the gyroscope (because of conservation of angular momentum) maintained its original orientation and appeared tilted relative to the locomotive. Today solid state technology replaces the rotating disk.

14. One pair of frequencies can be used at many locations if they are geographically separated to avoid interference. Six pairs of frequencies can cover the entire United States, assuming no more than six trains will simultaneously approach each other. More channels may be needed at certain locations.

15. "Deadline Extension Sought for System to Avoid Train Crashes," *Los Angeles Times*, March 17, 2011, accessed at articles.latimes.com.

16. Public Notice, Federal Communications Commission DA No. 11-838, May 5, 2011.

17. Other computerized systems are being developed to increase efficiency, such as precision dispatching, computerized scheduling, and computer control of the throttle (to name just a few), and are all proceeding independently of PTC.

Chapter 7. Moving at the Wrong Speed

1. Railroad curves in the United States are described as degrees of curvature. The degree of curvature is the angle defined by the two radiuses at the two ends of a 100-foot (30-m) arc along the curve. The radius of curvature is approximately 5,730 divided by the degree of curvature.

2. For those determined to sum their forces and equate them to m × a, a 150-lb (667-N) force is applied underneath the scale to accelerate the person up at 16 ft/sec^2 (4.87 m/s^2). This 150-lb (667-N) force is balanced by the 100-lb force (445-N) of gravity and the person's inertia resisting the acceleration.

3. There are numerous requirements for Class 8 track. For example, if a moving car slightly bounces, no single wheel may unload to less than 10% of its normal load. Also, testing is done at 10 mph (16 km/h) above the certified operating speed.

4. Figure 7.8 correctly shows schematically a frictional force occurring if the coupling is pushed or pulled. The draft gear sits in a yoke that slides back and forth to compress the draft gear whether the coupling is pushed or pulled. This is difficult to illustrate without several sketches that show the different movements.

Chapter 8. Bearing Failures

1. The Pennsylvania was the largest railroad company (by traffic and revenue) in America throughout most of its late nineteenth- and twentieth-century existence. In 1968, the Pennsylvania merged with its arch rival, the New York Central, and became the Penn Central. The Penn Central declared bankruptcy in 1970.

2. Moving 1 division of 15,000 infantry required 1,350 cars. An armored division with 3,000 vehicles required 75 trains of nearly 40 cars each. The railroads moved 90% of all army and navy equipment and 97% of all troops during World War II.

3. White, *The American Railroad Passenger Car*, 519.

4. The earth is roughly 24,000 miles (38,600 km) in circumference at the equator. It rotates 1 revolution every 24 hours, about 1,000 mph. There are no 1,000-mph (1,600-km/h) winds outside. The air adheres to the earth's surface and rotates with it.

5. A break-in period was once required for new cars. Better-manufactured surfaces have mostly eliminated engine break-in.

6. A transducer converts one form of energy into another. A light-emitting diode (LED) converts electrical energy into light energy. A piezoelectric crystal converts electrical energy into mechanical motion, creating the sound energy emanating from "ultrasonic" sensors used by engineers to examine metals for defects and doctors to examine patients.

7. Hanley and Klaveness, "Commuter Train Bound for Manhattan Derails, Injuring 13," 81.

8. In the late 1970s, the railroads refused to transport spent nuclear fuel; the federal courts ruled they must.

9. In a 12-year period between 1997 and 2008, there were 132 fatalities from shipping hazmat material by truck versus 18 from trains. In 2008, there were 14,781 hazmat trucking incidents reported versus 750 for trains.

10. Railroad companies often sponsor training of local firemen.

11. In the 1990s, 40 pressurized propane tanks were exposed to fire to study BLEVEs. Six containers were safely protected by their safety valves, 13 tanks exploded with a BLEVE. The rest ruptured with an ignited jet of pressurized propane blasting from the opening.

12. Hot box detectors are usually located near a siding so that trains can quickly stop for inspection—not so in this case. Also, the warning signal was not sent until the mile-long train completed passing the detector. After the accident, the detectors were reprogrammed to sound an immediate alarm.

13. If a loaded tank car can currently survive a 2.5-foot (0.76-m) vertical drop on its head, a 12-fold increase in crashworthiness allows the improved car to survive a drop $12 \times 2.5 = 30$ feet (9 m).

Chapter 9. Gravity

1. A world-class sprinter averages almost 23 mph (37 km/h) for 328 feet (100 m). An average person can perhaps run 15 mph (24 km/h) briefly. One has to imagine overweight, middle-age persons running after the train while slipping on the loose gravel track bed.

2. In 1995, 60 cars rolled 4,800 feet (1,463 m) on a 0.25% grade, a drop of just 3 inches (7.6 cm) every 100 feet (31 m). Wind conditions were not reported.

3. The definitive experiment was conducted in 1971, when a feather and a hammer hit the ground at the same time when dropped on the airless moon. On earth air resistance can cause significant variation in the speed of falling bodies.

4. Air pressure is determined by the weight of the column of air above the barometer. Higher altitude corresponds to a lower barometric reading.

5. Coupling cars once required the locomotive to gingerly nudge forward as a person, dangerously standing between the cars, inserted a pin in a link. With the introduction of automatic couplers at the end of the nineteenth century, the person was no longer needed and coupling occurred at higher speeds. Higher coupling speeds, poor-quality steel couplings, and more powerful locomotives made broken trains very common. One industry survey suggests nearly 100,000 broken trains in 1897 alone! After the train comes apart, the two parts can collide with each other. Broken train accidents rarely occur today.

6. A heavyweight Olympic-trained weightlifter can produce 6.4 horsepower (4,786 watts) for a fraction of a second. A normal person is comfortable with a sustained effort on a bicycle of 0.1 horsepower (75 watts).

7. If the diesel engine directly drove the powered axles (without a generator), very complicated and expensive gearing would be required. The diesel's power is distributed with electrical wires instead of mechanical gears.

8. NTSB report.

9. An embankment is earthen fill used to level out low areas. Embankments are used to minimize short up-and-down motion of the train that can create handling problems.

10. The erroneous weight, 6,150 tons, divided by the number of cars does not equal 69 tons per car. Since empty cars brake better than full cars, empty cars were counted as 1.5 cars for braking. Subsequent testing concluded the 1.5 multiplier was incorrect.

11. *San Francisco Chronicle*, May 13, 1989, A1.

12. *The New Yorker Magazine* reports political pressure to restart the pipeline and prevent gasoline shortages in Las Vegas.

13. The government studies disasters for a variety of planning purposes. One such reason is the possible future shipment by train of spent nuclear fuel rods to a site for permanent storage. Safe shipment depends on anticipating all worst-case scenarios. The nuclear containers, for example, have been designed for a head-on collision with a freight train traveling at 70 mph (113 km/h) and for various fire scenarios.

Chapter 10. More Runaways

1. Shaw, *Down Brakes*, 268.

2. "Senators Divided on Wrecks Cause," *New York Times*, January 23, 1953, 40.

3. Dirt or ice has been known to plug the 1.025-inch (3.8-cm)-diameter brake pipe. Dirt collectors make this unlikely but not impossible. The temperature was 53°F (11.6°C).

4. The brake cylinder's rod is not directly attached to a brake shoe, as shown in Figure 10.2, but instead activates the same linkages used by the hand brake.

5. Over the years attempts have been made to design gradual brake release in freight trains. A long freight train accumulates leakage along its length. Current rules allow up to 15 psi (103 kPa) lower pressure in the back of the train. A small pressure reduction corresponding to a gradual brake release could be confused with the small pressure reduction caused by leakage. Gradual release is accomplished in shorter passenger trains with a modified triple valve.

6. Westinghouse made his second fortune establishing alternating electric current (AC) as the standard instead of Edison's direct current (DC) system.

7. Standard DB is 10,000 lbs (44.5 kN) per axle of braking force, or 1 unit of DB. Just as computer traction control minimizes wheel slip and supplies more pulling tractive force (see Chapter 9), the same controls provide better braking torque. Depending on the traction control, some axles of DB have multipliers that raise the effective number of axles. Six axles of DB, with a multiplier of 1.5, count as 9 effective axles of DB.

8. Weights given are for a typical 36-inch (91-cm)-diameter freight car wheel.

9. Dynamic braking in the front during emergency braking can derail a train.

Emergency braking automatically sets the engines at idle and turns the dynamic brakes off.

10. Counting the locomotive's airbrakes, the investigators concluded that the train could safely descend at 15 mph (24 km/h).

11. There are other requirements as well. The insulation and protective jacket must meet strength requirements after aging at 250°F (121°C) for 4 days. The completed cable must also pass abrasion, penetration, and cold bend test and crush test (2,500 lbs [1,134 kg] on [11 kN] 2.25 inches [5.7 cm] of cable). All standards must be met after durability testing, including immersion in sand, liquids, and a salt spray. The insulation must still insulate while being soaked in water for 26 weeks and charged with 600 volts. One connector design remained free of dirt and corrosion after 2 years of testing but would absorb moisture, freeze, and not pull apart without damage. Back to the drawing board.

12. The first test trains in revenue service occurred in the United States in 1995. Revenue service began in Canada in 1998 and in South Africa in 2000.

13. During the mandated 1,000-mile (1,609-km) brake inspection, every brake shoe must be visually observed to properly engage and release. This could take hours for a 150-car freight train with $8 \times 150 = 1,200$ wheels and brakes.

14. Matthew L. Wald, "Amtrak Official Outlines Roots of Acela Problem," *New York Times*, May 12, 2005, A25.

Chapter 11. Broken Rail

1. *Daily Mail*, February 8, 2005.
2. Similar surface contact fatigue occurs in bearings and gears.

Chapter 12. Buckled Track

1. NTSB report.
2. *Times Picayune*, April 6, 2004, 1.
3. *Times Picayune*, April 6, 2004, 1.
4. *Florida Times-Union*, April 19, 2002, A1.
5. *Florida Times Union*, April 18, 2002, A1.
6. Steel of the same dimensions does not buckle. Aluminum has twice the coefficient of thermal expansion and one-third the stiffness of steel. This makes aluminum six times more liable to buckling than steel. A longer length and/or higher temperature is required to buckle a similar steel bar.

7. About 95% of all cross ties in the United States are made of wood. Concrete is used if there is trouble with maintenance (damp tunnels) or for critical applications (high-speed Amtrak trains). The heavier concrete increases track stability. Increasingly, concrete cross ties are being used on curves that require frequent adjustment.

8. When calculating the rail stretch, the weight of the rail is ignored for simplification.

9. To calculate the thermal force (for a given length of rail), first calculate the thermal expansion (or contraction) relative to its installation temperature. Strain equals the calculated thermal expansion divided by the length of rail. Recall that $30 \times 10^6 =$

stress divided by strain. Stress equals the unknown force being solved for divided by the cross sectional area of the rail. The unknown thermal force can now be solved for.

10. Rail is often manipulated to act as if it were installed at a higher temperature than ambient. This can be done by heating the rail above ambient or by hydraulically stretching the rail.

Epilogue. Safety in the Modern Era

1. Electronically Controlled Pneumatic (ECP) brakes with an electric wire running the length of the train to activate the airbrakes are currently in use on a few trains and expected to be a growing trend (see Chapter 9).

REFERENCES

Chapter 1. The Railroad Industry (as Seen through Accidents)

Adams, Charles Francis. "The Railroad Death-Rate." *Atlantic Monthly*, February 1876.

Aldrich, Mark. *Death Rode the Rails: American Railroad Accidents and Safety, 1825–1965*. Baltimore: Johns Hopkins University Press, 2006.

American Association of Railroads. "America's Freight Railroads: Global Leaders," April 2011.

American Association of Railroads. "The Impact of the Staggers Rail Act of 1980," September 2009.

Ellig, Jerry. "Railroad Deregulation and Consumer Welfare." *Journal of Regulatory Economics*, March 2002.

FRA. Charles W. McDonald. The Railroad Safety Program—100 Years of Safer Railroads, 1993.

FRA. Preliminary National Rail Plan. October 15, 2009.

Fronczak, Robert E. "U.S. Railroad Safety Statistics and Trends," paper prepared for the Association of American Railroads, Transportation External Coordination Working Group Meeting, September 21, 2005.

Hankey, John P., Peter A. Hansen, Bill Metzger, Bob Johnston, Michael W. Blaszak, Don Phillips, and David Lustig. "The Truth about Trains." *Trains*, July 2009.

"JR West Admits Fatal Crash Could Have Been Avoided." *Japan Times*, August 6, 2007.

Klein, Maury. "The Diesel Revolution." *Invention & Technology*, Winter 1991.

Prout, H. G. "Railroad Accidents in the United States and England." *North American Review*, University of Iowa, Cedar Falls, Iowa, December 1893.

"Railroad Regulation: Economic and Financial Impacts of the Staggers Rail Act of 1980," U.S. General Accounting Office Report to Congressional Requesters, May 1990.

Statistical Abstract of the United States, U.S. Department of Commerce, various years, 1925–2010.

Testimony by John H. Riley, Federal Railroad Administrator, before the Surface

Transportation Subcommittee of the Senate Committee on Commerce, Science, and Transportation, November 1, 1985.

Tran, Mark. "Indian Rail Boss Resigns." *The Guardian*, August 3, 1999.

U.S. Department of Transportation. "Commodity Flow Survey, Bureau of Transportation Statistics," 1997.

U.S. Department of Transportation. "Pocket Guide to Transportation, Bureau of Transportation Statistics," 1999–2010.

Vose, G. L. "Safety in Railway Travel." *North American Review*, University of Iowa, Cedar Falls, Iowa, December 1882.

Chapter 2. How Trains Crash, Then and Now

"100 Dead in the Paris Disaster." *New York Times*, August 12, 1903.

"121 Persons Are Killed and 57 Injured in Train Collision." *Nashville Tennessean*, July 9, 1918.

"180 DIE in Crash of Holiday Trains in Fog near Paris." *New York Times*, December 24, 1933.

"Accident at Quintinshill on 22 May 1915," Board of Trade, September, 17, 1915, www.railwaysarchive.co.uk/documents/BoT_Quin1915.pdf.

"All Safe in Wreck of Steel Car Train." *New York Times*, November 18, 1911.

Bellman, Eric. "As Mumbai Grows, Commuter Trains Turn Deadly." *Wall Street Journal*, April 19, 2007.

British Transport Commission. "Re-appraisal of the Plan for the Modernization and Re-equipment of British Railways," paper presented to Parliament by the Minister of Transport and Civil Aviation by Command of Her Majesty, July 1959.

California Public Utilities Commission, Consumer Protection & Safety Division, Consumer Rail Collision Avoidance Report R.08-11-017, Exhibit 6-A, "Cellular/Wireless Device Records Factual Report Metrolink Engineer," March 3, 2009.

Clark, Richard W. "Brief Overview of the Head-On Collision between Metrolink Train 111 and a Union Pacific Freight Train September 12, 2008," California Public Utilities Commission. "Metrolink Train 111" site, ca.gov.

Crashworthiness Factual Report—Addendum #1, National Transportation Safety Board, July 7, 2009.

Appendix A—Interview Transcript, Battalion Chief Norman Greengard

Appendix B—Interview Transcript, Assistant Chief Scott Mottram

Appendix C—Interview Transcript, Captain Steven Ruiz

Appendix D—Interview Transcript, Battalion Chief Joseph Castro

Appendix E—Interview Transcript, Captain Christopher Cooper

Appendix F—Interview Transcript, Captain Bill Bugg

Appendix G—Interview Transcript, Deputy Barry Ryan

Appendix H—Interview Transcript, Deputy Bill Lynch and Deputy Brad Johnson

Appendix I—Interview Transcript, Officer Jose Valle

Appendix J—Interview Transcript, Officer Timothy Wolleck and Officer Richard Moberg

Appendix K—Interview Transcript, Officer Samuel Hong and Deputy Probation Officer Percy Sanders

Appendix L—Interview Transcript, Battalion Chief Timothy Ernst

Appendix M—Interview Transcript, Battalion Chief Mark Jones

Appendix N—Interview Transcript, Sergeant Nina Sutter, Battalion Chief John Quintanar, and Jesus Ojeda

"Dead in Wreck 27." *New York Times*, September 1, 1943.

"Death Toll in Train Blast Is 18." *Montgomery News*, January 3, 1935.

"Derailed in Big Storm." *New York Times*, June 17, 1925.

Editors of Consumer Reports. *Consumer Reports Buying Guide 2009.*

Engel, Mary, and Rich Connell. "Metrolink Collision: Learning from the Past." *Los Angeles Times*, September 14, 2008.

"Engine in Wreck Explodes." *New York Times*, November 9, 1905.

FRA. Emergency Order to Restrict On-Duty Railroad Operating Employees' Use of Cellular Telephones and Other Distracting Electronic and Electrical Devices, October 7, 2008.

FRA. Evaluation of Concepts for Locomotive Crew Egress, DOT/FRA/ORD-03/07, March 2003.

FRA. Federal Track Safety Standards Fact Sheet, Office of Public Affairs, June 2008.

FRA. Fire Safety of Passenger Trains: Phase II: Application of Fire Hazard Analysis Techniques, DOT/FRA/ORD-01/16, Washington, D.C., December 2001.

FRA. Quick-Release Emergency Egress Panels for Cab Car End Door, RR08-28, November 2008.

French, Peter W. "U.S. Railroad Safety Statistics and Trends: AVP-Safety & Performance Analysis," Association of American Railroads, July 29, 2008.

Gold, Scott, and Molly Hennessy-Fiske. "Metrolink Collision." *Los Angeles Times*, September 19, 2008.

Interstate Commerce Commission. "Investigation of an Accident Which Occurred on the Nashville, Chattanooga & St. Louis Railway at Nashville, Tennessee, on July 9, 1918." Submitted August 16, 1918.

Judge, Tom. "How Does Your Track Measure Up?" *Railway Age*, September 2004.

Lopez, Robert J., Rich Connell, and Steve Hymon. "Train Engineer Sent Text Message Just Before Crash." *Los Angeles Times*, October 2, 2008.

Lopez, Robert J., Garrett Therolf, and Scott Gold. "Metrolink Collision: A Massive Response." *Los Angeles Times*, September 14, 2008.

Miller, Barbara. "Mosaic of Errors Caused Alpine Ski-Train Fire That Killed 155." *The Independent*, June 19, 2002.

Ministry of Transport Railroad Accidents. Report on the Collision which occurred on 2nd December 1955 near Barnes Station on the Southern Region British Railways, June 27, 1956.

Ministry of Transport Railroad Accidents. Report on the Fire which occurred in an Express Passenger Train on 14th July 1951 near Huntingdon in the Eastern Region British Railways, London, 1952.

Moger, Robin. "Train Inferno Kills 373 Passengers." *The Telegraph*, February 21, 2002, www.telegraph.co.uk/news/worldnews/africaandindianocean/egypt/1385592/Train-inferno-kills-373-passengers.html.

NTSB. 20594 Docket No. DCA-08-MR009. Washington, D.C.

NTSB. Collision of Metrolink Train 111 with Union Pacific Train LOF65-12, Chatsworth, California, September 12, 2008. Railroad Accident Report NTSB/RAR-10/01. Adopted January 21, 2010.

NTSB. Railroad Accident Report, Southern Pacific Transportation Co., Freight Train 2nd BSM 22 Munitions Explosion, Benson, Arizona, May 24, 1973, NTSB-RAR-75-2. Adopted on February 26, 1975.

Pierson, David, Scott Glover, and Scott Gold. "Metrolink Collision." Los Angeles Times, September 13, 2008.

Reed, Robert C. Train Wrecks. New York: Bonanza Books, 1968.

Resnikoff, Marvin. "Study of Transportation Accident Severity Radioactive Waste Management," Nevada Nuclear Waste Project Office, February 1992.

Rubin, Joel, Ann M. Simmons, and Mitchell Landsberg. "Metrolink Crash: Carnage in Chatsworth." Los Angeles Times, September 13, 2008.

"Safety in Steel Coaches." New York Times, September 10, 1913.

Semmens, Peter. Railway Disasters of the World. London: Patrick Stephens Limited, 1994.

"Seven Die in Flames of Wrecked Train." New York Times, February 23, 1909.

Shaw, Robert B. Down Brakes. London: P. R. MacMillan Limited, 1961.

"A Train Becomes a Fiery 'Tunnel of Death.'" Maclean's, March 4, 2002.

"Transport: Wrecks." Time, January 7, 1935.

Tucker, Farrell L. "The Great Locomotive Explosion: A Socio-Historical Examination of a Tragedy," www.colfa.utsa.edu/users/jreynolds/Tucker/expl.html. Accessed February 27, 2010.

White, John H., Jr. The American Railroad Freight Car. Baltimore: Johns Hopkins University Press, 1995.

White, John H., Jr. The American Railroad Passenger Car. Baltimore: Johns Hopkins University Press, 1978.

"Wreck Kills Five; Four Trains Crash." New York Times, June 8, 1911.

"Wreck on Boston & Albany." New York Times, December 15, 1907.

"Wreck Toll Laid to Wooden Cars." New York Times, December 26, 1933.

Chapter 3. Moderate-Speed Passenger Train Collisions

"4 Killed in Crash on the New Haven." New York Times, August 21, 1969.

Bibel, George. Beyond the Black Box: The Forensics of Airplane Crashes. Baltimore: Johns Hopkins University Press, 2008.

Bowen, Douglas John. "Multiple-unit of Locomotive Hauled?" Railway Age, June 2009.

Carolan, M., B. A. Perlman, and D. Tyrell. "Evaluation of Occupant Volume Strength in Conventional Passenger Railroad Equipment." Proceedings of RTDF2008, 2008 ASME Rail Transportation Division Fall Technical Conference, September 24–25, 2008, Chicago, Illinois.

Carolan, Michael, and Michelle Priante Muhlanger. "Strategy for Alternative Occupant Volume Testing." Proceedings of ASME 2009, Rail Transportation Division Fall Conference RTDF 2009, October 19–21, 2009, Fort Worth, Texas.

"Cars Telescoped." *New York Times*, November 23, 1950.

Code of Federal Regulations Title 49: Transportation; PART 238—PASSENGER EQUIPMENT SAFETY STANDARDS; Subpart E—Specific Requirements for Tier II Passenger Equipment.

Connell, Rich. "Metrolink's Collapsible Trains." *Los Angeles Times*, May 2, 2010.

"Derailment at Eschede" ("High Speed Train Wreck"). Seconds From Disaster. Episode 6, Season 1. First aired on National Geographic Channel, August 3, 2004.

FRA. Investigation Report HQ-2005-0, Southern California Regional Rail Authority (SCRX), Glendale, California, January 26, 2005.

FRA. Office of Safety Office of Railroad Development, Report to the House and Senate Appropriations Committees: The Safety Of Push-Pull and Multiple-Unit Locomotive Passenger Rail Operations, Washington, D.C., June 2006.

FRA. Passenger Rail Train-to-Train Impact Test Volume II: Summary of Occupant Protection Program.

FRA. U.S. Department of Transportation, Federal Railroad Administration Office of Research and Development. DOT/FRA/ORD-03/17.II, Washington, D.C., July 2003.

ICC. Ex Parte No. 176, Accident near Jamaica on November 22, 1950, N.Y. Submitted November 28, 1950.

ICC. Investigation No. 2856, The Southern Pacific Company, Accident near Bagley, Utah, on December 31, 1944. Adopted March 7, 1945.

ICC. Report of the Director, Bureau of Safety, Accident on the Southern Pacific Railroad, Tortuga, California, September 20, 1938, Investigation No. 2294, Washington, 1939.

ICC. Report No. 3349, The Long Island Railroad Company in Re Accident at Huntington, N.Y., on August 5, 1950. Washington, D.C. Adopted October 4, 1950.

Jacobsen, Karina, Kristine Severson, and Benjamin Perlman. "Effectiveness of Alternative Rail Passenger Equipment Crashworthiness Strategies," paper presented at the 2006 ASME/IEEE Joint Rail Conference, April 4–6, 2006, Atlanta, Georgia.

James, Gerard. Analysis of Traffic Load Effects on Railway Bridges, Structural Engineering Division, Royal Institute of Technology, Stockholm, Sweden, TRITA-BKN, Bulletin 70, 2003.

Liu, Caitlin. "Testimony of Train Crash Witness Calls Suicide Theory into Question." *Los Angeles Times*, May 4, 2005.

MacNeill, Robert A., and Steven W. Kirkpatrick. "Vehicle Postmortem and Data Analysis of a Passenger Rail Car Collision Test." Proceedings of JRC 2002, 2002 ASME/IEEE Joint Rail Conference, April 23–25, 2002, Washington, D.C.

Martinez, Eloy, et al. "Crush Analyses of Multi-Level Equipment." Proceedings of IMECE2006, 2006 ASME International Mechanical Engineering Congress and Exposition, November 5–10, 2006, Chicago, Illinois.

Martinez, Eloy, David Tyrell, Robert Rancatore, and Richard Stringfellow. "A Crush Zone Design for An Existing Passenger Rail Cab Car." Proceedings of 2005 ASME International Mechanical Engineering Congress and Exposition, November 5–11, 2005, Orlando, Florida.

Mayville, R. A., R. G. Stringfellow, R. J. Rancatore, and T. P. Hosmer. Locomotive Crashworthiness Research, Volume 1: Model Development and Validation, DOT/FRA/ORD-95/08.1, 1995.

"Metrolink Gets on the Safe Track: Crash-absorbing Rail Cars Unveiled." Daily News, May 4, 2010.

NTSB. Accident Report NTSB-RAR-82-1, Beverly, Massachusetts, August 11, 1981. Washington, D.C. Adopted March 9, 1982.

NTSB. Collision between Two Washington Metropolitan Area Transit Authority Trains at the Woodley Park-Zoo / Adams Morgan Station in Washington, D.C., November 3, 2004, National Transportation Safety Board, Washington, D.C., NTSB/RAR-06/01. Adopted March 23, 2006.

NTSB. Collision of Burlington Northern Santa Fe Freight Train with Metrolink Passenger Train, Placentia, California, April 23, 2002. Washington, D.C. Adopted October 7, 2003.

NTSB. Collision of Illinois Central Gulf Railroad Commuter Trains, Chicago, Illinois, October 30, 1972. Washington, D.C. Adopted April 25, 1973.

NTSB. Collision of Metrolink Train 111 with Union Pacific Train LOF65-12, Chatsworth, California, September 12, 2008. Washington, D.C. Adopted January 21, 2010.

NTSB. Collision of National Railroad Passenger Corporation (Amtrak) Train 59 with a Loaded Truck-Semitrailer Combination at a Highway/Rail Grade Crossing in Bourbonnais, Illinois, March 15, 1999. Washington, D.C. Adopted February 5, 2002.

NTSB. Collision of Northern Indiana Commuter Transportation District Train 102 with a Tractor-Trailer, Portage, Indiana, June 18, 1998. Washington, D.C. Adopted August 10, 1999.

NTSB. Collision of Trains N-48 and N-49 at Darien, Connecticut, August 20, 1969. Washington D.C. Adopted October 14, 1970.

NTSB. Collision of Two Washington Metropolitan Area Transit Authority Trains near Fort Totten Station, Washington, D.C., June 22, 2009, Public Docket, Panel 1, Exhibit P1-h Crashworthiness of WMATA Fleet, "Washington Metropolitan Area Transit Authority Railcar Crashworthiness," February 18, 2010.

NTSB. Collision of Washington Metropolitan Area Transit Authority Train T-111 with Standing Train at Shady Grove Passenger Station, Gaithersburg, Maryland, January 6, 1996. Washington, D.C. Adopted October 29, 1996.

NTSB. Collision of WMATA Metrorail Train 112 with WMATA Metrorail Train 214, near the Fort Totten Station in the District of Columbia (DC), on June 22, 2000, Crashworthiness Factual Report, December 15, 2009.

NTSB. Derailment of Steam Excursion Train Norfolk and Western Railway Company

Train, Extra 611 West, Suffolk, Virginia, May 18, 1986. Washington, D.C. Adopted January 12, 1987.

NTSB. Head-End Collision of Amtrak Train No. 392 and Ice Train No. 51, Harvey, Illinois, October 12, 1979. Washington, D.C. Adopted April 3, 1980.

NTSB. Head-On Collision of Chicago, South Shore and South Bend Railroad Trains Nos. 123 and 218, Gary, Indiana, January 21, 1985, Accident Report RAR-85-13. Washington, D.C. Adopted October 21, 1985.

NTSB. Head-On Collision of National Railroad Passenger Corporation (Amtrak); Passenger Trains Nos. 151 and 168, Astoria, Queens, New York, New York, July 23, 1984. Railroad Accident Report NTSB/RAR-85/09. Washington, D.C. Adopted May 14, 1985.

NTSB. Safety Recommendation, R-87-30 and 31, Washington, D.C., December 9, 1987.

NTSB. Safety Recommendation, Washington, D.C., letter from Mark V. Rosenker, December 20, 2007.

O'Driscoll, Patrick. "Rail Safety Test a Smashing Success." *USA Today*, March 24, 2006.

Parent, D., D. Tyrell, and A. B. Perlman. "Crashworthiness Analysis of the Placentia, CA Rail Collision." Proceedings of ICrash 2004, International Crashworthiness Conference, July 14–16, 2004, San Francisco, California.

Priante, Michelle, and Eloy Martinez. "Crash Energy Management Crush Zone Designs: Features, Functions, and Forms." Proceedings of the ASME/IEEE Joint Rail Conference and Internal Combustion Engine Spring Technical Conference, March 13–16, 2007, Pueblo, Colorado.

Priante, M., D. Tyrell, and A. B. Perlman. "A Collision Dynamics Model of a Multi-Level Train." Proceedings of IMECE2006, 2006 ASME International Mechanical Engineering Congress and Exposition, Paper No. IMECE2006-13537, November 5–10, 2006.

Queensland Transport. Independent Support for the Cairns Tilt Train Accident Investigation; Stage 1—Design Compliance Matrix, Interfleet Technology Pty Ltd, Report No. ITPLR/1767/01, Sydney, Australia, April 3, 2009.

Rail Vehicle Crashworthiness Symposium, June 24–26, 1996, U.S. Department of Transportation Research and Special Programs Administration, Volpe National Transportation Systems Center, DOT/FRA/ORD-97/08, March 1998, Cambridge, Massachusetts.

Severson, K., and D. Parent. "Train-to-Train Impact Test of Crash Energy Management Passenger Rail Equipment: Occupant Experiments." Proceedings of IMECE2006, 2006 ASME International Mechanical Engineering Congress and Exposition, November 5–10, 2006, Chicago, Illinois.

Severson, Kristine J., David C. Tyrell, and A. Benjamin Perlman. "Collision Safety Comparison of Conventional and Crash Energy Management Passenger Rail Car Designs." Proceedings of ASME/IEEE Joint Railroad Conference, April 22–24, 2003.

Simmons, Ann M. "Metrolink Killer Gets 11 Life Terms, No Parole." *Los Angeles Times*, August 21, 2008.

Stringfellow, R., and C. Paetsch. "Modeling Material Failure During Cab Car End Frame Impact," American Society of Mechanical Engineers, Paper No. JRC2009-63054, March 2009.

Strong, Phil. "Multiple-unit vs. Locomotive Hauled." Letters to the Editor. *Railway Age*, August 2009.

Sun, Lena H. "Newer Metro Rail Cars to Buffer Oldest in Fleet." *The Washington Post*, June 30, 2009.

Tyrell, D. "Rail Passenger Equipment Accidents and the Evaluation of Crashworthiness Strategies," paper presented at the Rail Equipment Crashworthiness Symposium, Institute of Mechanical Engineers, May 2, 2001, London, England.

Tyrell, David, et al. "Overview of a Crash Energy Management Specification for Passenger Rail Equipment." Proceedings of Joint Rail Conference, April 4–6, 2006, Atlanta, Georgia.

Tyrell, David, et al. Passenger Rail Two-Car Impact Test Volume I: Overview and Selected Results, U.S. Department of Transportation, Federal Railroad Administration Office of Research and Development, DOT/FRA/ORD-01/22.I, Washington, D.C., January 2002.

Tyrell, David, Karina Jacobsen, and Eloy Martinez. "A Train-to-Train Impact Test of Crash Energy Management Passenger Rail Equipment: Structural Results." Proceedings of 2006 ASME International Mechanical Engineering Congress and Exposition, November 5–10, 2006, Chicago, Illinois.

Tyrell, D., K. Severson, A. B. Perlman, B. Brickle, and C. VanIngen-Dunn. "Rail Passenger Equipment Crashworthiness Testing Requirements and Implementation," *Rail Transportation*, American Society of Mechanical Engineers, RTD-Vol. 19, 2000.

Tyrell, D., K. Severson, A. B. Perlman, and R. Rancatore. "Train-to-Train Impact Test: Analysis of Structural Measurements," American Society of Mechanical Engineers, Paper No. IMECE2002-33247, November 2002.

Tyrell, D., K. Severson, J. Zolock, and A. B. Perlman. Passenger Rail Two-Car Impact Test Volume I: Overview and Selected Results, U.S. Department of Transportation, DOT/FRA/ORD-01/22.I, January 2002.

Tyrell, D., J. Zolock, and C. VanIngen-Dunn. "Train-to-Train Impact Test: Occupant Protection Experiments," American Society of Mechanical Engineers, Paper No. IMECE2002-39611, November 2002.

VanIngen-Dunn, C. Passenger Rail Train-to-Train Impact Test Volume II: Summary of Occupant Protection Program, U.S. Department of Transportation, DOT/FRA/ORD-03/17.II, July 2003.

Vantuono, William C. "Amtrak's Vision: Today, the Northeast. Tomorrow, America." *Railway Age*, April 1999.

Weir, Kytja. "Metro Eyes Purchase of New, Yet Incompatible, Rail Cars." *The Examiner*, March 26, 2010.

Chapter 4. Freight Train Collisions

Federal Register, Part II Department of Transportation, Federal Railroad Administration. 49 CFR Parts 229 and 238; Locomotive Crashworthiness; Final Rule, June 28, 2006.

FRA. Accident/Incident Investigation Report HQ-2005-33.

FRA. Accident/Incident Investigation Report HQ-2005-78.

FRA. Accident/Incident Investigation Report HQ-2005-97.

FRA. Accident/Incident Investigation Report HQ-2006-04. January 18, 2006— Lincoln, Alabama—Rear End Collision—Norfolk Southern (NS).

FRA. Accident/Incident Investigation Report HQ-2006-30. May 17, 2006—Lake Side, Nebraska—Rear End Collision—Burlington Northern Santa Fe (BNSF).

FRA. Accident/Incident Investigation Report HQ-2007-72. November 10, 2007— Niland, California—Rear End Collision—Union Pacific (UP).

FRA. Collision Analysis Working Group (CAWG), 65 Main-Track Train Collisions, 1997 through 2002: Review, Analysis, Findings, and Recommendations, CAWG Final Report, July 2006.

FRA. Crashworthiness Design Modifications for Locomotive and Cab Car Anticlimbing Systems, DOT/FRA/ORD-03/05.

FRA. Locomotive Crashworthiness Research: Modeling, Simulation, and Validation, DOT/FRA/ORD-01/23.

FRA. Office of Safety Headquarters Assigned Accident Investigation Report HQ-2005-54.

FRA. Railroad Accident Investigation Report No. 4105, Terminal Railroad Association of St. Louis, St. Louis, Missouri, November 18, 1966. Washington, D.C. Adopted April 18, 1967.

FRA. Railroad Accident Investigation Report No. 4158, Penn Central Company, Wellington, Ohio, August 18, 1969. Washington, D.C. Adopted September 24, 1970.

FRA. Railroad Accident Investigation Report No. 4160, Southern Railway Company, October 26, 1969. Washington, D.C. Adopted November 9, 1970.

FRA. Railroad Accident Investigation Report No. 4168, Chicago, Rock Island and Pacific Railroad Company, Union, Missouri, March 25, 1970. Washington, D.C. Adopted June 24, 1971.

FRA. Railroad Accident Investigation Report No. 4185, The Atchison, Topeka and Santa Fe Railway Company, Duncanville, Texas. Washington, D.C. Adopted June 9, 1972.

Locomotive Crashworthiness Research Volume 1: Model Development and Validation, DOT/FRA/ORD-95/08.1, Volpe National Transportation Systems, Cambridge, MA 02142-1093, June 1995.

Locomotive Crashworthiness Research Volume 4: Additional Freight Locomotive Calculations, DOT/FRA/ORD-95/08.4, Volpe National Transportation Systems, Cambridge, MA 02142-1093, July 1995.

NTSB. CHI 97 FR 014, Collision/Fire/Employee Fatality, Union Pacific Railroad, Fort Worth, Texas, August 20, 1997. No adoption date.

NTSB. Collision between Two BNSF Railway Company Freight Trains near Gunter, Texas, May 19, 2004, Railroad Accident Report NTSB/RAR-06/02. Washington, D.C. Adopted June 13, 2006.

NTSB. Collision Involving Three Consolidated Rail Corporation Freight Trains Operating in Fog on a Double Main Track near Bryan, Ohio, January 17, 1999, NTSB/RAR-01/01. Washington, D.C. Adopted May 9, 2001.

NTSB. Collision of Burlington Northern Santa Fe Railway in Scottsbluff, Nebraska, February 13, 2003. Washington, D.C. Adopted February 13, 2004.

NTSB. Collision of Norfolk Southern Corporation Train 255L5 with Consolidated Rail Corporation Train TV 220 in Butler, Indiana, on March 25, 1998, Railroad Accident Report NTSB/RAR-99/02, Washington, D.C.

NTSB. Collision of Norfolk Southern Freight Train 192 with Standing Norfolk Southern Local Train P22, South Carolina, on January 6, 2005. Washington, D.C. Adopted November 29, 2005.

NTSB. Collision of Two Burlington Northern Santa Fe Freight Trains near Clarendon, Texas, May 28, 2002, Railroad Accident Report NTSB/RAR-03/01. Washington, D.C. Adopted June 3, 2003.

NTSB. Collision of Two Canadian National/Illinois Central Railway Trains near Clarkston, Michigan, November 15, 2001. Washington, D.C. Adopted November 19, 2002.

NTSB. Collision of Two CN Freight Trains, Anding, Mississippi, July 10, 2005. Washington, D.C. Adopted March 20, 2007.

NTSB. Collision of Union Pacific Railroad Train MHOTU-23 with BNSF Railway Company Train MEAP-TUL-126-D with Subsequent Derailment and Hazardous Materials Release, Macdona, Texas, June 28, 2004. Adopted July 6, 2006.

NTSB. Crashworthiness Factual Report. Rear End Collision on April 17, 2011. Filed on September 12, 2011.

NTSB. Head-On Collision of Boston & Maine Corporation, Extra 1731 East and Massachusetts Bay Transportation Authority Train No. 570 on Former Boston & Maine Corporation Tracks, Beverly, Massachusetts, August 11, 1981, NTSB-RAR-82-1. Adopted March 9, 1982.

NTSB. Illinois Central Railroad Company and Indiana Harbor Belt Railroad Company Collision between Yard Trains at Riverdale, Illinois, on September 8, 1970.

NTSB. Railroad Accident Brief, Accident No. DCA-04-FR-006, Carrizozo, New Mexico, February 21, 2004. Adopted October 31, 2006.

NTSB. Railroad Accident Brief, Accident No. DCA-04-MR-003, Kelso, Washington, on November 15, 2003, Burlington Northern Santa Fe Railway Company and Union Pacific Railroad. Washington, D.C. Adopted June 6, 2005.

NTSB. Railroad Accident Brief Report ATL 97 FR 005, Rear-End Collision, Union Pacific Railroad Company, Odem, Texas, February 21, 1997. Washington, D.C. Adopted April 23, 1998.

NTSB. Railroad Accident Brief Report, ATL97FR020, Rear-End Collision/Derailment Conrail, Hummelstown, Pennsylvania, September 29, 1997. Washington, D.C. Adopted December 1, 1998.

NTSB. Railroad Accident Brief, Collision of Two Union Pacific Railroad Trains, Shepherd, Texas, NTSB/RAB-06/01, September 15, 2005. Adopted May 22, 2006.

NTSB. Railroad Accident Brief, Rear-End Collision and Collision with Derailed Equipment, DCA-02-FR-002, Missouri, December 13, 2001. Adopted June 17, 2004.

NTSB. Railroad Accident Brief, Rear-end Collision of Norfolk Southern Trains near Lincoln, Alabama, January 18, 2006. Washington, D.C. Adopted October 26, 2007.

NTSB. Railroad Accident Brief Report, LAX 96 FR 006, Head-On Collision, Union Pacific Railroad, Nacco, Wyoming, January 31, 1996. Washington, D.C. Adopted August 18, 1998.

NTSB. Safety Recommendation H-82-1 through 3. Issued March 25, 1982.

NTSB. Special Study: Railroad/Highway Grade Crossing Accidents Involving Trucks Transporting Bulk Hazardous Materials, Washington, D.C., September 24, 1981.

Punwani, Swamidas, Gopal Samavedam, and Steve Kokkins. "Advances in the Railroad Locomotive Crashworthiness," paper presented at the 2003 ASME International Mechanical Engineering Congress, November 15–21, 2003, Washington, D.C.

Transportation Safety Board of Canada. Collision and Derailment on July 2, 1996. Report Number R96W0171. Adopted on April 9, 1999.

Transportation Safety Board of Canada. Collision on August 12, 1996. Report Number R96C0172. Adopted August 22, 1997.

Transportation Safety Board of Canada. Collision on February 22, 2002. Report Number R02T0047. Adopted on July 16, 2003.

Transportation Safety Board of Canada. Collision on January 31, 1999. Report Number R99E0023. Adopted on August 21, 2001.

Transportation Safety Board of Canada. Derailment/Collision on April 23, 1999. Report Number R99H0007. Adopted February 13, 2001.

Chapter 5. Avoiding Collisions

"47 Die in Wreck, 100 Hurt in Illinois." *New York Times*, April 26, 1946.

"100 Years of Subway Signals." *Railway Age*, June 2004.

Aldrich, Mark. "Combating the Collision Horror: The Interstate Commerce Commission and Automatic Train Control, 1900–1939." *Technology and Culture*, January 1993.

Aldrich, Mark. *Death Rode the Rails: American Railroad Accidents and Safety, 1825–1965*. Baltimore: Johns Hopkins University Press, 2006.

"Amtrak to Add Wi-Fi to Some Trains." *USA Today*, January 14, 2010.

"Automatic Train Stops." *New York Times*, May 14, 1924.

Bernstein, Victor. "More and More Speed by Rail." *New York Times*, November 1, 1936.

Blaszak, Michael W. "Speed, Signals and Safety." *Fast Trains*, 2009.

Bonney, Joe. "Untangling the Chicago Knot." *Journal of Commerce*, April 20, 2009.

Bryan, Frank W. "CTC: Remotely Directing the Movement of Trains." *Trains*, May 1, 2006.

Burlington Northern Santa Fe Corporation. 2007 Annual Report.

Chicago "L".org, www.chicago-l.org/operations/towers/wilson.html.

Clark, Thomas. *The American Railway, Its Construction, Development, Management and Appliances.* New York: Charles Scribner & Sons, 1889.

Coel, Margaret. "A Silver Streak." *Invention & Technology*, Fall 1986.

Corliss, Carlton Jonathan. "The Day of Two Noons." Association of American Railroads, Washington, D.C., 1942.

"Crash Comes at Crossing." *New York Times*, February 28, 1921.

"Deaths Put at 44 in Illinois Wreck." *New York Times*, April 27, 1946.

de Cerreño, Allison L. C., and Shishir Mathur. Mineta MTI Report 06-03, High-Speed Rail Projects in the United States: Identifying the Elements of Success, Part 2, Transportation Institute, San Jose, California, November 2006.

Dunlap, David W. "Clearing the Tracks for Penn Station III." *New York Times*, January 3, 1999.

"Engine Crew Held for Porter Wreck." *New York Times*, March 5, 1921.

FRA. Accident Investigation Report HQ-2005-33, Union Pacific, Blairstown, Iowa, April 13, 2005.

FRA. Accident Investigation Report HQ-2005-97, CSX Transportation, Mauk, Georgia, October 31, 2005.

FRA. Accident Investigation Report HQ-2006-04, January 18, 2006—Lincoln, Alabama—Rear End Collision—Norfolk Southern (NS).

FRA. Accident Investigation Report HQ-2006-30, May 17, 2006—Lake Side, Nebraska—Rear End Collision—Burlington Northern Santa Fe (BNSF).

FRA. Collision Analysis Working Group (CAWG), 65 Main-Track Train Collisions, 1997 through 2002: Review, Analysis, Findings, and Recommendations, July 2006.

FRA. D. B. Devoe, An Analysis of the Job of Railroad Train Dispatcher. Washington, D.C., April 1974.

FRA. Judith Gertler and David Nash. Optimizing Staffing Levels and Schedules for Railroad Dispatching Centers, DOT/FRA/ORD-04/01, Washington, D.C., September 2004.

FRA. Kenneth B. Ullman and Alan J. Bing, High Speed Passenger Trains in Freight Railroad Corridors: Operations and Safety Considerations, DOT/FRA/ORD-95/05, Washington, D.C., 1994.

FRA. Office of Safety Analysis, wwwsafetydata.fra.dot.gov/OfficeofSafety/publicsite/Query/gxrtally1.aspx.

FRA. "Positive Train Control Overview," www.fra.dot.gov/us/content/1265.

FRA. Press release, June 12, 2008 (Washington, D.C.), www.fra.dot.gov/us/press-releases/188.

FRA. Project Development and Selection Process, www.fra.dot.gov/us/content/888.

FRA. Railroad Research and Development Program, www.fra.dot.gov/us/content/1241.

FRA. Report to the House and Senate Appropriations Committees: The Safety of Push-Pull and Multiple-Unit Locomotive Passenger Rail Operations, Washington, D.C., June 2006.

FRA. Signals and Train Control Fact Sheet, Office of Public Affairs, Washington, D.C., October 2008.

FRA. Train Dispatchers Follow-Up Review, Report to Congress, Washington, D.C., January 1995.

General Railway Signal Company. *Electric Interlocking Handbook*. Rochester, N.Y., 1913.

Gnaedinger, Louis B. N. "Rail Gains Marked by Streamlining." *New York Times*, December 31, 1934.

Grynbaum, Michael M. "The Zoo That Is Grand Central, at Full Gallop." *New York Times*, November 25, 2009.

Gwyer, William L. "Train Orders." *Trains*, May 1, 2006.

ICC. Railroad Accident Investigation, Ex Porte No. 215, The Central Railroad Company of New Jersey, Elizabethport, N.J., September 15, 1958. Submitted October 31, 1958.

"Illinois Wreck Could Have Been Avoided." *New York Times*, May 3, 1946.

Institution of Railway Signal Engineers. *Introduction to North American Railway Signaling*. Omaha, Neb.: Simmon-Boardman Books, 2008.

Kaempffert, Waldemar. "The Dawn of Railroad Transformation." *New York Times*, July 9, 1933.

King, Everett Edgar. *Railway Signaling*. New York: McGraw Hill, 1921.

Kos, Sayre C., and Andy Cummings. "Ask Trains." *Trains*, June 2007.

Kos, Sayre C., and Jon Roma. "Ask Trains." *Trains*, February 2008.

"Machinery Takes the Place of Man." *New York Times*, March 28, 1909.

McDonald, Charles W. The Federal Railroad Safety Program, 100 Years of Safer Railroads, August 1993, www.fra.dot.gov/downloads/safety/rail_safety_program_booklet_v2.pdf.

"Michigan Central Train Runs Over Derail Onto Crossing and New York Central Train Crashes into It." *Railway Age*, March 4, 1921.

Middleton, William D. *When the Steam Railroad Electrified*. 2nd ed. Bloomington: Indiana University Press, 2001.

Middleton, William D., and Mark Reutter. "Fast Trains and Faster." *Railroad History*, Spring–Summer 2007.

Nash, Andrew. "Best Practices in Shared-Use High-Speed Rail Systems." MTI Report 02-02, Mineta Transportation Institute, San José State University. June 2003.

NTSB. Collision between Union Pacific Freight Trains MKSNP-01 and ZSEME-29 near Delia, Kansas, July 2, 1997. Washington, D.C. Adopted August 31, 1999.

NTSB. Collision of Illinois Central Gulf Railroad Commuter Trains, Chicago, Illinois, October 30, 1972. Washington, D.C. Adopted June 28, 1973.

NTSB. Collision of Two Burlington Northern Santa Fe Freight Trains near Clarendon, Texas, May 28, 2002. Washington, D.C. Adopted June 3, 2003.

NTSB. Crashworthiness Factual Report, Collision of SCRRA ("Metrolink") Train 111 with Union Pacific Train LOF65-12, in the Chatsworth District of the City of Los Angeles, California, on September 12, 2008. Washington, D.C. Submitted February 20, 2009.

NTSB. DCA-08-MR-009; Metrolink-UP Collision, Chatsworth, CA, September 12, 2008, Signal Group Factual Report from the Public Docket, January 20, 2009.

NTSB. Railroad Accident Report NTSB/RAR-10/02. Washington, D.C. Adopted July 27, 2010.

NTSB. Recommendations and Accomplishments, Most Wanted List, Transportation Safety Improvement, Railroad, Implement Positive Train Control Systems, Washington, D.C., November 2007.

NTSB. Survival Factors Group Chairman's Factual Report from the Public Docket, Emergency Response, January 30, 2009.

"One Signal Tower Controls Seventy-Nine Acres of Tracks." *New York Times*, February 2, 1913.

"Orders Safety Devices." *New York Times*, June 19, 1947.

Pennsylvania Railroad, Hunter Interlocking Tower. Historic American Engineering Record, National Park Service, Philadelphia, Pennsylvania (undated). Posted at Library of Congress. www.loc.gov/pictures/item/NJ1745/.

Peterman, David Randall. "Amtrak: Overview and Options," Congressional Research Service, Library of Congress, Washington, D.C., January 25, 2001.

Resor, Randolph R., et al. "Positive Train Control (PTC): Calculating Benefits and Costs of a New Railroad Control Technology." *Journal of the Transportation Research Forum*, Summer 2005.

Reutter, Mark. "The Lost Promise of the American Railroad." *Wilson Quarterly*, Winter 1994.

Roma, Jon R. "City of Towers." *Trains*, July 2003.

Schneider, Paul D. "The Double-Track Dilemma." *Trains*, July 1991.

Seth, S. King. "44 Die, 320 Hurt as Chicago Commuter Train Crashes." *New York Times*, October 31, 1972.

Shaw, Robert B. *A History of Railroad Accidents, Safety Precautions and Operating Practices in the United States of America*. Binghamton, N.Y.: Vail-Ballou Press, 1978.

Showalter, William Joseph. "America's Amazing Railway Traffic." *National Geographic Magazine*, April 1923.

Smedley, Steve. "Last 'Armstrong' Tower Falls." *Trains*, August 2010.

Solomon, Brian. *Railroad Signaling*. St. Paul, Minn.: MBI Publishing Company, 2003.

Stern, Jessica. "Resignaling Grand Central." *Railway Age*, January 1992.

"Train Crash Laid to Speed." *New York Times*, March 1, 1925.

"Trains Here Hold Records for Speed." *New York Times*, April 5, 1936.

"Two Tracks to Do the Work of Four." *New York Times*, January 17, 1957.

Ullman, Kenneth B., and Alan J. Bing. High-Speed Passenger Trains in Freight Train Corridors: Operations and Safety Considerations, FRA/ORD-95/05, Washington, D.C., 1994.

Vine, Ken, and Phil Hingley. "Sustainable Interlocking Technology for the 21st Century," paper presented at the IRSE Aspect 2003 Conference, September 2003, London, England.

Wald, Matthew. "Amtrak's Wariness Imperils Grand Central-L.I.R.R. Link." *New York Times*, February 1, 2004.

"What It Means to Run a Great Railroad." *New York Times*, March 31, 1907.

White, M. P. "Sustainable Interlocking for the 21st Century." *Institution of Railway Signal Engineers (IRSE) in UK Monthly Newsletter*, September 2007.

Wreathall J., D. Bley, E. Roth, J. Multer, and T. Raslear. "Using an Integrated Process of Data and Modeling in HRA." *Reliability Engineering and System Safety*, February 2004.

"Zephyr Makes World Record Run." *New York Times*, May 27, 1934.

Chapter 6. Positive Train Control

"2d Sentence Is Imposed in 1987 Amtrak Crash." *New York Times*, July 19, 1988.

215 Million People Rode Metrorail in Fiscal Year 2008, Metro press release, July 8, 2008, www.wmata.com/about_metro/news/PressReleaseDetail.cfm?Release ID=2179.

Alibrahim, Sam. Signal and Train Control, Federal Railroad Administration Research & Development Program Overview, Washington, D.C., November 13–14, 2008.

Allen, Leonard W., III, Program Manager, Office of Research and Development, Federal Railroad Administration. "Intelligent Railroad Systems," paper presented at the Civil GPS Service Interface Committee meeting, September 12, 2005, Washington, D.C.

Association of American Railroads. "Positive Train Control," June 2010.

Baugher, Roger W. "PTC: Overlay, or Stand-Alone?" *Railway Age*, May 2004.

Bureau of Transportation Statistics, U.S. Department of Transportation. *Pocket Guide to Transportation 2009*. Washington, D.C.

Commuter Rail Collision Avoidance Report, Railroad Operations Safety Branch of the Consumer Protection & Safety Division, San Francisco, California, December 15, 2009.

Conkey, Christopher. "Safety Costs Chafe Railroad." *Wall Street Journal*, October 26, 2009.

Connel, Rich. "Deadline Extension Sought for System to Avoid Train Crashes." *Los Angeles Times*, March 17, 2011.

Cunningham, Joseph. "Roots of an Evolution." *Railway Age*, April 2009.

Divis, Dee Ann. "Augmented GPS: All in the Family." *GPS World*, April 1, 2003.

Electronic Train Management System (ETMS), Waiver Hearing—Testimony of United Transportation Union Paul Thompson, Rick Marceau, and James Stem to the Federal Railroad Association, February 23, 2006.

FCC. Public Notice DA No. 11-838, May 5, 2011, Washington, D.C.

FCC. Staff Report on NTIA's Study of Current and Future Spectrum Use by the Energy, Water and Railroad Industries, Federal Communications Commission, July 30, 2002.

FRA. 49 CFR Parts 229, 234, 235, and 236 [Docket No. FRA-2008-0132, Notice No. 3], Positive Train Control Systems. Washington, D.C. Issued December 30, 2009.

FRA. 49 CFR Part 229, 234, 235, et al., Positive Train Control Systems, Final Rule, Washington, D.C., January 15, 2010.

FRA. Development of an Adaptive Predictive Braking Enforcement Algorithm, Washington, D.C., April 2009.

FRA. Development of a General Train Movement Simulator for Safety Evaluation, Washington, D.C., February 2007.

FRA. ECP Brake System for Freight Service, Final Report, August 2006.

FRA. Human Reliability Analysis in Support of Risk Assessment for Positive Train Control, DOT/FRA/ORD-03/15, Washington, D.C., June 2003.

FRA. A Practical Risk Assessment Methodology for Safety-Critical Train Control Systems, DOT/FRA/ORD-09/15, Washington, D.C., July 2009.

FRA. Safety of Highway-Railroad Grade Crossings; Research Needs Workshop, Volume I, DOT/FRA/ORD-95/14.1, Washington, D.C., January 1996.

Frailey, Fred W. "Biggest Technical Challenge Railroads Ever Faced." *Trains*, April 2011.

Frailey, Fred W. "Rails Prepare to Test PTC." *Trains*, April 2011.

"GE Transportation Introduces Connection Family of Positive Train Control Products at Rail Exhibition." *Mass Transit*, May 22, 2009.

Hansen, Peter A. "Railroads Vow to Make PTC Work." *Trains*, January 2002.

Heckler, D. A. "Challenges for Railroad Radio." *Urgent Communications*, May 1, 2000.

Hilkevitch, Jon. "Amtrak Able to Travel to 110 mph in Indiana, Michigan." *Chicago Tribune*, February 7, 2012.

Hoelsher, James, and Larry Light. "Full PTC Today with Off the Shelf Technology: Amtrak's ACSES Overlay on Expanded ATC." Proceedings of the AREMA 2001 Annual Conferences, September 9–12, 2001, Chicago, Illinois.

Hohmann, James. "Still Seeking a Cause, Investigators Conduct Simulation at Accident Site." *The Washington Post*, July 19, 2009.

Jordon, Steve. "U.P.: System Stops Crashes." *Omaha World-Herald*, January 23, 1987.

Light, P. E., and E. Lawrence. "Practical Positive Train Control in the Northeast Corridor and in an Emerging Corridors," paper presented at the Transportation Research Board 86th Annual Meeting, 2007.

Lindsey, Ron. "Is PTC a True Market Driver?" *Railway Age*, July 2007.

Memorandum to Secretarial Officers Modal Administrators from Joel Szabat, Assistant Secretary for Transportation Policy, Office of the Secretary of Transportation, Washington, D.C., March 18, 2009.

Metro Facts. www.wmata.com/about_metro/docs/metrofacts.pdf.

Metrolink Quarterly Report, Operations Committee, Los Angeles, California, May 19, 2011.

Metrolink Quarterly Report, Operations Committee, September 15, 2011.

Mitchell, Alexander D. "BNSF Tests Positive Train Control in Illinois." *Trains*, May 2006.

Moody, Howard, Edwin F. Kemp, and Robert Gallamore. "Radio: The Next Wave—Radio Usage in Railroads." *Railway Age*, October 1999.

NDGPS. Assessment Final Report, Federal Highway Administration, McLean, Virginia, March 2006.

NTSB. Collision of Washington Metropolitan Area Transit Authority Train T-1 11 with Standing Train at Shady Grove Passenger Station, Gaithersburg, Maryland, January 6, 1996. Washington, D.C. Adopted October 29, 1996.

NTSB. Derailment of Washington Metropolitan Area Transit Authority Train near the Mt. Vernon Square Station, Washington, D.C., January 7, 2007, NTSB/RAR-07/03. Washington, D.C. Adopted October 16, 2007.

NTSB. Rear-End Collision of Amtrak Passenger Train 94, The Colonial, and Consolidated Rail Corporation, Freight Train Ens-121, on the Northeast Corridor, Chase, Maryland, January 4, 1987, NTSB/RAR-88/01. Washington, D.C. Adopted January 1988.

Petit, Bill. "PTC Closes in on Interoperability." *Railway Age*, May 2008.

Petit, Bill. "Vital—What Does it Really Mean?" *Railway Systems and Controls*, September 2005.

Petit, William A. "Interoperable Positive Train Control (PTC)," paper presented at the AREMA 2009 Annual Conference.

Petition of Association of American Railroads (AAR) for Modification of Licenses for Use in Advanced Train Control Systems and Positive Train Control Systems. Adopted February 13, 2001, before the Federal Communications Commission, Washington, D.C.

Polivka, Alan, et al. North American Joint Positive Train Control Project, DOT/FRA/ORD-09/04, FRA, Washington, D.C., April 2009.

"Positive Train Control—Ready to Go?" *Mass Transit*, December 17, 2007.

"PTC System Passes Critical Test." *Railway Age*, January 2003.

Rail Safety: Federal Railroad Administration Should Report on Risks to the Successful Implementation of Mandated Safety Technology, GAO-11-133, December 2010.

Starcic, Janna. "Positive Train Control Implementation Signals Challenges Ahead." *Metro Magazine*, August 2010.

Statement of Joseph H. Boardman, Administrator, Federal Railroad Administration, U.S. Department of Transportation before Senator Barbara Boxer, U.S. Senate, September 23, 2008.

Stephens, Joe, and Lena H. Sun. "Metro Scare under Potomac." *The Washington Post*, September 6, 2009.

Sun, Lena H. "Metro Told NTSB about Safety System Malfunction." *The Washington Post*, August 13, 2009.

Sun, Lena H., and Maria Glod. "Metro Control System Fails Test." *The Washington Post*, June 26, 2009.

Sun, Lena H., and Maria Glod. "Probe Finds Metro Control 'Anomalies.'" *The Washington Post*, June 25, 2009.

Sun, Lena H., and Lyndsey Layton. "Circuit Problem in Metro Crash Began in '07, NTSB Says." *The Washington Post*, July 24, 2009.

Sun, Lena H., and Lyndsey Layton. "Metro Failed to Detect Hazard." *The Washington Post*, July 2, 2009.

Title 49—Transportation; Chapter II—Federal Railroad Administration, Department of Transportation; Part 236—Rules, Standards, and Instructions Governing the Installation, Inspection, Maintenance, and Repair of Signal and Train Control Systems, Devices, and Appliances; Appendix C to Part 236—Safety Assurance Criteria and Processes.

Tse, Terry. Federal Railroad Administration Research and Development Program Review, Highlights of R&D Activities for PTC Implementation, Signals, Train Control and Communications Division, Washington, D.C., March 12, 2009.

Vantuono, William C. "Traffic Control Gets Smarter." *Railway Age*, May 2007.

Weir, Kytja. "Metro Says It Can't Yet Comply." *The Examiner*, July 14, 2009.

Wilner, Frank N. "STB Expert: Consider PTC's Business Benefits." *Railway Age*, November 2008.

Chapter 7. Moving at the Wrong Speed

Aldrich, Mark. *Death Rode the Rails: American Railroad Accidents and Safety, 1825–1965*. Baltimore: Johns Hopkins University Press, 2006.

Barkan, Christopher, P. L. Dick, C. Tyler, and Robert Anderson. "Railroad Derailment Factors Affecting Hazardous Materials Transportation Risk." *Transportation Research Record*, 2003.

Barrow, Keith. "The Long and the Short of Distributed Power." *Railway Age*, August 2011.

Code of Federal Regulations. Title 49—Transportation; Chapter II—Federal Railroad Administration, Sections 213.57, 213.329, 213.333, and 238.427.

"Conductor Asserts He Felt No Braking Before Rail Wreck." *New York Times*, February 12, 1951.

Dawson, Richard W. "Making the Connection." *Trains*, August 2000.

El-Sibaie, Magdy A. Coupler Angling under In-Train Loads: Modeling and Validation, Association of American Railroads, R-772, Pueblo, Colorado, March 1991.

FRA. Accident Investigation Report HQ-2005-10, February 1, 2005.

FRA. Accident Investigation Report HQ-2005-21, March 6, 2005.

FRA. Accident Investigation Report HQ-2006-10, February 10, 2006.

FRA. Accident Investigation Report HQ-2006-92, November 29, 2006.

FRA. Accident Investigation Report HQ-2006-101, December 16, 2006.

FRA. Accident Investigation Report HQ-2007-35, June 10, 2007.

FRA. Advanced Truck for Higher-Speed Freight Operations, RR 09-07, April 2009.

FRA. E. H. Law et al., General Models for Lateral Stability Analysis of Railway Freight Vehicles, FRA/ORD-77/36, June 1977.

FRA. Impact Performance of Draft Gears in 263,000 Pound Gross Rail Load and 286,000 Pound Gross Rail Load Tank Car Service, DOT/FRA/ORD-06/16, June 2006.

FRA. Safe Placement of Cars, Report to the Senate Committee on Commerce, Science, and Transportation and the House Committee on Transportation and Infrastructure, June 2005.

"Grim Workers Toil at Scene of Wreck." *New York Times*, February 8, 1951.

Hay, William W. *Railroad Engineering*. New York: John Wiley & Sons, 1953.

ICC. Accident at Woodbridge, N.J., on February 6, 1951. Adopted April 19, 1951.

ICC. Report No. 3675, The Atchison, Topeka and Santa Fe Railway Company, Accident at Redondo Jct., Los Angeles, California, on January 22, 1956. Adopted March 15, 1956.

"Jersey Rail Wreck Laid to High Speed and Lack of Warning Signal at Span." *New York Times*, February 8, 1951.

Johnston, Bob. "Behind the Changes in Amtrak's New Timetable." *Trains*, December 2004.

Judge, Tom. "Beyond 1 Million Miles." *Railway Age*, June 2002.

King, Leo. "Acela Woes Recede." *Destination Freedom Newsletter*, August 19, 2002.

"Long Trains and Brakes." *Reflexions*, Winter 2004.

Luczak, Marybeth. "In It for the Long Haul." *Railway Age*, March 2004.

Lustig, David. "Engineers to Get Computer Help." *Train*, May 2007.

Machalaba, Daniel. "Amtrak's Temporary Fixes Aim to Restore Most Acela Service." *Wall Street Journal*, August 19, 2002.

Martin, G. C., and W. W. Hay. Method of Analysis for Determining the Coupler Forces and Longitudinal Motion of a Long Freight Train in Over-The-Road Operation, Railway Research Department of Civil Engineering, University of Illinois, Urbana, Illinois, June 1967.

Middleton, William D., et al., eds. *Encyclopedia of North American Railroads*. Bloomington: Indiana University Press, 2007.

"Night Scene Grim." *New York Times*, February 7, 1951.

Norfolk Southern. Locomotive Engineer Training Handbook. February 2006.

NTSB. Derailment of Northeast Illinois Regional Commuter Railroad Train 519 in Chicago, Illinois, October 12, 2003. Adopted November 16, 2005.

NTSB. Railroad Accident Brief, DCA-05-MR-013, Chicago, Illinois, September 17, 2005. Adopted December 21, 2006.

NTSB. Railroad Accident Brief Report, ATL 97 FR 011, Derailment Consolidated Rail Corporation, Sandusky, Ohio, April 21, 1997. Adopted August 18, 1998.

NTSB. Railroad Accident Report NTSB-RAR-79-4, Derailment of Southern Railway Company Train No. 2, The Crescent, Elma, Virginia, December 3, 1978. Adopted June 7, 1979.

Parola, John. "Systems of Track Infrastructure Safety at Amtrak," paper presented at the AREMA 2001 Conference.

"Passengers Tell of Swaying Cars." *New York Times*, February 7, 1951.

Sawley, Kevin. "Refining the Wheel/Rail Interface." *Railway Age*, April 2002.

Schneider, Paul D. "Growing Pains for the Biggest Locomotives." *Trains*, August 2001.

Sharke, Paul. "No Hunting." *Mechanical Engineering*, May 2001.

Sherrock, Eric, et al. "High Speed Interface." *Railway Age*, June 2008.

Tournay, Harry M., Scott Cummings, and Ron Lang. "A New Twist on Bad Actors." *Railway Age*, February 2006.

Transportation Safety Board of Canada. Derailment on August 5, 2005. Report Number R05V0141. Adopted May 30, 2007.

Transportation Safety Board of Canada. Derailment on February 7, 2004. Report Number R04Q0006. Adopted on January 10, 2006.

Transportation Safety Board of Canada. Derailment on February 12, 2007. Report Number R07D0009. Adopted July 30, 2008.

Transportation Safety Board of Canada. Derailment on January 8, 2001. Report Number R01W0007. Adopted on February 12, 2003.

Transportation Safety Board of Canada. Derailment on May 16, 2000. Report Number R00W0106. Adopted on April 12, 2002.

Transportation Safety Board of Canada. Derailment on October 6, 2001. Report Number R01M0061. Adopted on January 8, 2004.

Transportation Safety Board of Canada. Derailment/Collision on April 23, 1999. Report Number R99H0007. Adopted February 13, 2001.

Transportation Safety Board of Canada. Investigation of Yard Impact Forces on 286,000 lb Tank Cars, Transport Canada Report TP 13591E, March 2000.

TTCI 15th Annual AAR Research Review, Pueblo, Colorado, March 2–3, 2010.

Vantuono, William C. "Climbing the Pumpkin Vine." *Railway Age*, July 2007.

Vantuono, William C. "Control This!" *Railway Age*, April 2002.

Vantuono, William C. "New Choices for Better Ride Quality." *Railway Age*, May 1999.

Wheelihan, Jack. "Helpers." *Trains*, August 2001.

Wolf, Gary. "It Takes Three to Rock and Roll." *Rail Sciences Inc*, February 2005.

Ytuarte, Christopher. "Ingredients for a Smooth Ride." *Railway Age*, December 2002.

Chapter 8. Bearing Failures

"5th Anniversary of Derailment Mississauga Crash: The Fears Remain." *The Globe and Mail*, November 10, 1984.

"200,000 in Mississauga and Oakville Moved from Homes as Chlorine Escapes Tank Cars Ablaze after Derailment." *The Globe and Mail*, November 12, 1979.

Adams, Frank S. "More Than 50 Are Killed in Wreck of Speeding Congressional Limited in the Outskirts of Philadelphia." *New York Times*, September 7, 1943.

Aldrich, Mark. *Death Rode the Rails: American Railroad Accidents and Safety, 1825–1965*. Baltimore: John Hopkins University Press, 2006.

"Are We Really Serious About Hot Boxes?" *Railway Age*, February 11, 1952.

Asimov, Isaac. *Understanding Physics, 3 Volumes in One*. New York: Dorset Press, 1988.

Bruce, J. Plain. "Bearing for Railway Wagons." *Tribology International*, October 1975.

Chiles, James R. "Small Wonder: The Magnificent Ball Bearing." *American Heritage*, Summer 1985.

Code of Federal Regulations. Title 49, 179.14, "Coupler Vertical Restraint System."

English, G. W., and T. W. Moynihan. "Causes of Accidents and Mitigation Strategies,

Submitted to Railway Safety Act Review Secretariat, Kingston Ontario," July 2007.

FRA. Thomas V. Peacock and Hugh H. Snider Jr., Railroad Car Roller Bearing Temperature Measurement and Analysis, FRA/ORD-80/43, April 1980.

FRA. Roller Bearing Failure Mechanisms Test and Wheel Anomaly Test Report, DOT/FRA/ORD-92/08, June 1992.

Gerard, Warren. "Mississauga Nightmare." *Maclean's*, November 26, 1979.

Green, D. F., et al. "Recent Developments in Plain Bearings." *Railway Engineering Journal*, 1975.

Griswold, Wesley S. *Train Wreck*. Battleboro, Vt.: Stephen Greene Press, 1969.

Hanley, Robert, and Charles Klaveness. "Commuter Train Bound for Manhattan Derails, Injuring 13." *New York Times*, July 15, 2003.

Hay, William W. *Railroad Engineering*. New York: John Wiley & Sons, 1953.

Hazardous Materials: Improving the Safety of Railroad Tank Car Transportation of Hazardous Materials, Final Rule, U.S. Department of Transportation, January 13, 2009.

Hildebrand, Michael S., and Gregory G. Noll. *Propane Emergencies*. 3rd ed. Chester, Md.: Red Hat Publishing Co., 2007.

Horger, Oscar. "If All Freight Cars had Roller Bearings (Part I)." *Railway Age*, December 31, 1951.

"Hot Box Detected Too Late on P. R. R." *New York Times*, September 10, 1943.

ICC. Investigation No. 2726, The Pennsylvania Railroad Company Report In Re Accident At Shore Pa., on September 6, 1943.

ICC. Investigation No. 2856, The Southern Pacific Company Accident near Bagley, Utah, on December 31, 1944.

ICC. Investigation No. 2921, In the Matter of Making Accident Investigation Reports Under the Accident Reports Act of May 6, 1910, Great Northern Railway Company. Adopted September 20, 1945.

ICC. Report of the Chief of the Bureau of Safety Covering the Investigation of an Accident Which Occurred on the Michigan Central Railroad at Ivanhoe, Ind., on June 22, 1918. Adopted August 8, 1918.

"Journal Bearing Load Service Environment—Freight Car Tests." *Railway Age*, December 1991.

Kube, Kathi. "Improving Hazmat Transportation." *Trains*, January 2010.

Lamb, J. Parker. *Perfecting the American Steam Locomotive*. Bloomington: Indiana University Press, 2003.

Lawrence, R., et al. "Analysis of Journal Roller Bearing Failure at Fast on May 29, 1986," paper presented at the ASME Winter Annual Meeting, December 1987, Boston, Massachusetts.

Leedham, Robert C., and William N. Weins. "Mechanistic Aspects of Bearing Burnoff," paper presented at the ASME Winter Annual Meeting, 1992.

Leibensperger, Robert. "The Conquest of Friction." *Mechanical Engineering*, November 2003.

Le Massena, Robert A. "Timken vs. Everyone Else." *Trains*, November 1997.

Maintenance of Railway Cars. Department of the Army Technical Manual TM 55-203, Washington, D.C., August 1972.

McCooey, Paula. "Fiery Rail Crash Forces 500 to Flee." *The Ottawa Citizen*, February 22, 2003.

McGrew, J. M., et al. "Reliability of Railroad Roller Bearings." *Journal of Lubrication Technology*, January 1977.

Metals Handbook, Volume 11. *Failure Analysis and Prevention*. 9th ed. American Society for Metals, 1986.

Michael, C., et al. "Railroad Journal Roller Bearing Failure and Detection Past, Present and Future," paper presented at the ASME Winter Annual Meeting, December 16, 1987, Boston, Massachusetts.

Mississauga Train Derailment (1979), www.mississauga.ca/portal/residents/local history?paf_gear_id=9700018&itemId=5500001.

Morris, John D. "Railroads Show Strain of Peak War Traffic." *New York Times*, September 12, 1943.

"Moving Troops Has Become a Bigger Job." *Railway Age*, August 22, 1942.

Muendel, John. "Friction and Lubrication in Medieval Europe." *Isis*, September 1995.

Norris, Craig. "Failure Progression Modes (FPM): A Newly Proposed Bearing Failure Mode Identification and Classification System," paper presented at Mechanical Association Railcar Technical Services, 44th Annual Technical Conference, September 19–21, 2005, Chicago, Illinois.

NTSB. Recent Accident History of Hot Box Detector Data Management, NTSB-SIR-81-1, Washington, D.C., 1981.

"Report of Committee on Rails." *Railway Age*, March 21, 1942.

"Roller Bearings Breach the Last Barrier." *Railway Age*, February 8, 1947.

Semmens, Peter. *Railway Disasters of the World*. Somerset, England: Haynes Publishing, 1994.

Shigley, Joseph, et al. *Mechanical Engineering Design*. 5th ed. New York: McGraw Hill, 1989.

Statement of Thomas D. Simpson before the U.S. House of Representatives Committee on Transportation and Infrastructure Subcommittee on Railroads, June 13, 2006.

Sullivan, Richard. "Panel Discussion & Series of Papers: Freight Car Wheels & Roller Bearings," paper presented at Mechanical Association Railcar Technical Services, 45th Annual Technical Conference, September 19–21, 2006, Chicago, Illinois.

Transportation Safety Board of Canada. Derailment on February 21, 2003. Report Number R03T0080. Adopted on July 27, 2004.

Transportation Safety Board of Canada. Derailment on May 21, 2003. Report Number R03T0158. Adopted October 12, 2004.

Transportation Safety Board of Canada. Derailment on October 19, 2003. Report Number R03W0169. Adopted July 7, 2004.

Transportation Safety Board of Canada. Derailment/Collision on February 21, 2003. Report Number R03T0080. Adopted July 27, 2004.

U.S. Department of Transportation, Bureau of Transportation Statistics. National Transportation Statistics, 2009, Research and Innovative Technology.

Vantuono, William C. "Brutish Bearings Bear the Brunt." *Railway Age*, February 2003.

Vantuono, William C. "It's What's Inside That Counts." *Railway Age*, February 2008.

Walker, Russell, and Gerald Anderson. "Sounding Out Those 'Growlers.'" *Railway Age*, May 2007.

Wang, Hao. "Axle–Burn Off Stack Up Force Analysis of a Railroad Roller Bearing Using the Finite Element Method." Ph.D. thesis, University of Illinois, 1995.

Welty, Gus. "When Bearings Fail . . . —Railroad Cars." *Railway Age*, December 1991.

"The Wheel That Got Away." *Railway Age*, August 2003.

Williams, W. M., et al. "Postmortem Metallurgical, Thermal and Mechanical Analysis of Railroad Journal Burnoff," paper presented at the ASME Conference, December 13–18, 1987, Boston, Massachusetts.

Wolf, Gary. "Derailment Prevention: Early Detection Is the Key." *Railway Age*, February 1998.

Written Statement of Joseph H. Boardman before the U.S. House of Representatives Committee on Transportation & Infrastructure Subcommittee on Railroads, June 13, 2006.

Chapter 9. Gravity

"3 Killed as Runaway Train Smashes Into Row of Houses." *San Francisco Chronicle*, May 13, 1989.

Accident Returns: Extract for Accident at Armagh on 12 June 1889, Board of Trade (Railway Department), London, July 8, 1889.

Aldrich, Mark. *Death Rode the Rails: American Railroad Accidents and Safety, 1825–1965*. Baltimore: Johns Hopkins University Press, 2006.

Amsted Rail, Technical Forum, Bearing Rolling Resistance, www.amstedrail.com/tech_sheets/0301.asp.

"Around the World, 196 Died When Stones Used as Brake Failed to Hold Train." *Chicago Tribune*, May 27, 2002.

Asimov, Isaac. *Understanding Physics: Motion, Sound and Heat*. New York: Signet Classic, 1969.

Beard, Matthew. "Tsunami Disaster: Briton Among Few to Survive on 'Queen of the Sea' Train." *The Independent*, December 31, 2004.

Berton, Pierre. *The Impossible Railway*. New York: Knopf, 1972.

Blaine, D. G. Determining Practical Tonnage Limits and Speeds in Grade Operations, ASME-Paper 69-WA/RR-6, 1969.

Blaine, D. G., G. M. Cabble, and F. J. Grejda. Combination Friction Braking Systems for Freight Cars, American Society of Mechanical Engineers (Paper), 75-RT-11, 1975.

Blaine, D. G., F. J. Grejda, and J. C. Kahr. Braking Duty in North American Freight Train Service and Effects on Brake Equipment, Brake Shoes and Wheels. American Society of Mechanical Engineers (Paper), 78-RT-9, 1978.

Blaine, D. G., and M. F. Hengel. Brake-system Operation and Testing Procedures and Their Effects on Train Performance, ASME Paper 71-WA/RT-9, 1971.

Blaine, D. G., M. F. Hengel, and J. H. Peterson. "Train Brake and Track Capacity Requirements for the '80s." *Journal of Engineering for Industry, Transactions of the ASME*, November 1981.

Brennan, Pat. Orange County Register, Santa Anna, California, June 10, 1989.

Clodfelter, Frank. "Saluda." *Trains*, November 1984.

Dempsey, Hugh A., ed. *The CPR West*. Toronto: Douglas & McIntyre, 1984.

"Excess Load Cited in Train Accident." *New York Times*, May 18, 1989.

Faith, Nicholas. *Derail: Why Trains Crash*. London: Macmillan Publishers, 2000.

FRA. Comparative Safety of the Transport of High-Level Radioactive Materials on Dedicated, Key, and Regular Trains, DOT/FRA/ORD-05/03, Washington, D.C., March 2006.

FRA. Resistance of a Freight Train to Forward Motion—Volume I, Methodology & Evaluation, FRA/ORD 78/04.I, April 1978.

Gibbons, John Murray. *The Romantic History of the Canadian Pacific*. Toronto: McClelland & Stewart, 1935.

Graham, Hatch, and Judy Graham. "The Great San Bernardino Train Wreck." *Dog Sports*, September 1989.

Hankey, John P., et al. "The Truth about Trains." *Trains*, July 2009.

Hemphill, Mark W. "Ask Trains." *Trains*, October 2001.

Jones, H. H., and D. G. Blaine. "Railway Freight Train Braking—North American Style," *Institution of Mechanical Engineers, Conference Publications*, 1979.

"Killed on an Excursion." *New York Times*, June 13, 1889.

King, Ed. "Up the Grade the Norfolk & Western Way." *Trains*, April 2004.

Loehne, Bob. "Saluda Grade Cheats Death." *Trains*, November 1994.

Marks, Lionel, ed. *Mechanical Engineers Handbook*. 3rd ed. New York: McGraw-Hill, 1941.

McGonigal, Robert S. "Grades and Curves." *Trains*, May 1, 2006.

"The Mountain Way." *Trains*, April 2004.

Mydans, Seth. "Runaway Freight Train Derails Killing 3." *New York Times*, May 13, 1989.

Norfolk Southern. Locomotive Engineer Training Handbook. February 2006.

NTSB. Accident Brief, LAX 96 FR 006 Head-On Collision, Union Pacific Railroad, Nacco, Wyoming, January 31, 1996. Adopted August 18, 1998.

NTSB. City of Commerce, California, Derailment on June 20, 2003, Accident Number DCA-03-FR-005. Adopted April 2004.

NTSB. Railroad Accident Report, Collision between Two Washington Metropolitan Area Transit Authority Trains at the Woodley Park-Zoo / Adams Morgan Station in Washington, D.C., November 3, 2004.

NTSB. Railroad Accident Report—Derailment of Southern Pacific Transportation Company Freight Train on May 12, 1989 and Subsequent Rupture of Calnev Petroleum Pipeline on May 25, 1989—San Bernardino, California, NTSB/RA-90/06. Adopted on June 19, 1990.

"Officials Question Train's Load Weight." *Houston Chronicle*, May 14, 1989.

"Opportunity Missed to Prevent Runaway Train's Derailment." *Houston Chronicle*, June 22, 2003.

"Pipeline Blast Kills 3 at Site of Derailment." *Chicago Sun Times*, May 26, 1989.

Potter, Jay. "The Standard: B&O's West End." *Trains*, April 2004.

"Screams, Prayers Filled Air as Train Rolled Backward." *Chicago Tribune*, June 26, 2002.

Semmens, Peter. *Railroad Disasters of the World*. London: Patrick Stephens Limited, 1994.

Shaw, Robert B. *Down Brakes*. London: P. R. MacMillan Limited, 1961.

"Train Crash Toll Is 281." *New York Times*, June 28, 2002.

"Train Derailment Warning Isn't Required by Law." *Daily Breeze*, June 29, 2003.

Transportation Safety Board of Canada. Derailment on March 21, 2009. Report Number R09T0092. Adopted September 20, 2010.

Transportation Safety Board of Canada. Runaway on April 13, 1996. Report Number R96C0086. Adopted May 1998.

Transportation Safety Board of Canada. Runaway on August 27, 1999. Report Number R99D0159. Adopted May 2001.

Transportation Safety Board of Canada. Runaway on December 14, 1995. Report Number R95M0072. Adopted April 1997.

Transportation Safety Board of Canada. Runaway/Derailment on December 2, 1997. Report Number R97C0147. Adopted December 1999.

Trow, George W. S. "Devastation." *New Yorker*, October 22, 1990.

Chapter 10. More Runaways

"3rd Braking Incident Is Checked by FBI." *New York Times*, January 16, 1953.

"The AAR Advanced Braking System—Association of American Railroads' Research and Test Department Investigation into Technological Innovations in Freight Railroad Braking Systems." *Railway Age*, March 1994.

AAR Specification S-4210. Performance Specification for ECP Brake System Cable, Connectors and Junction Boxes. Adopted May 1997.

The "AB" Freight Brake Equipment Instruction Pamphlet No. 5062. Wilmerding, Pa.: Westinghouse Air Brake Company, 1945.

The American Public Transportation Association. APTA RP-E-014-99, Recommended Practice for Diesel Electric Passenger Locomotive Blended Brake Control. Approved March 4, 1999; reaffirmed June 15, 2006.

Blaine, David G. "Determining Practical Tonnage Limits and Speeds in Grade Operations," American Society of Mechanical Engineers, 69-WA/RR-6, 1969.

Blaine, David G. *Modern Freight Car Air Brakes*. Omaha, Neb.: Simmon Boardman, 1979.

Blaine, D. G., M. F. Hengel, and J. H. Peterson. "Train Brake and Track Capacity Requirements for the '80s." *Journal of Engineering for Industry, Transactions of the ASME*, November 1981.

BNSF. Air Brake and Train Handling Rules, July 2003.

"The Burlington Brake Trials." *Railroad Gazette*, May 13, 1887.

Carlson, Fred, and Brian Smith. "ECP Braking Gets Results." *Railway Age*, October 2000.

Code of Federal Regulations. Title 49: Transportation; PART 232—Brake System Safety Standards for Freight and Other Non-Passenger Trains and Equipment, End-of-Train Devices, 232.109.

Cummings, Andy. "NS Begins Use of ECP Brakes." *Trains*, January 2008.

De Steese, John G., et al. "Electric Power from Ambient Energy Sources," PNNL-13336, paper prepared for the U.S. Department of Energy by Pacific Northwest National Laboratory, Richland, Washington, September 2000.

Driesbach, Douglas A., and Gus Welty. "A.C. The Switch Is On—Rail Transit Systems Moving to A.C. Power." *Railway Age*, August 1989.

Editors of Consumer Reports. *Consumer Reports Buying Guide 2009*.

"Electro-pneumatic Brakes." *Railway Age*, September 1999.

FRA. 49 CFR Parts 223, 229, 232, and 238, Passenger Equipment Safety Standards, Proposed Rule June 17, 1996.

FRA. ECP Brake System for Freight Service, August 2006.

FRA. Investigation of Cracks in Acela Coach Car Brake Discs: Test and Analysis Volume I—Final Report, DOT/FRA/ORD-06/07.1, Washington D.C., November 30, 2005.

FRA. Safety of High Speed Ground Transportation Systems: Safety of Advanced Braking Concepts for High Speed Ground Transportation Systems, DOT/FRA/ORD-95/09.1, September 1995.

FRA. Title 49—Transportation, Chapter II—Federal Railroad Administration, Department of Transportation. Part 232, Brake System Safety Standards for Freight and Other Non-Subpart B, General Requirements, Sec. 232.109 Dynamic brake requirements.

FRA. Track Compliance Manual, Chapter 6, Track Safety Standards Classes 6 through 9, January 1, 2002.

Furlong, Jim. "Amtrak Acela Problems Wheel into Past." *Destination: Freedom; The Newsletter of the National Corridors Initiative*, January 27, 2003.

Garhammer, J. "Power Production by Olympic Weightlifters." *Medicine and Science in Sports and Exercise*, Spring 1980.

"George Westinghouse." *New York Times*, March 13, 1914.

Gurstelle, William. *Backyard Ballistics*. Chicago: Review Press, 2001.

Holliday, R., and P. Goodman. "Going for Gold." *IEE Review*, May 2002.

ICC. File No. 3497-A, Ex Parte No. 184, Accident at Union Station, Washington, D.C., February 17, 1953.

Kearney, Jay T. "Training the Olympic Athlete." *Scientific American*, June 1996.

Klein, Maury. "The Diesel Revolution." *Invention & Technology*, Winter 1991.

Knowles, Clayton. "Senators Divided on Wreck's Cause." *New York Times*, January 23, 1953.

Lustig, David. "What Does Dynamic Braking Do?" *Trains*, July 2007.

Machalaba, Daniel, and Christopher J. Chipello. "Amtrak Scrutinizes Design Change." *Wall Street Journal*, June 28, 2005.

National Railroad Passenger Corporation. Amtrak, Air Brake and Train Handling Rules and Instructions, August 19, 2002.

Norfolk Southern. Locomotive Engineer Training Handbook. February 2006.

NTSB. Derailment of Freight Train H-Balti-31, Atchison, Topeka and Santa Fe Railway Company, near Cajun Junction, California, on February 1, 1996, NTSB/RAR-96/05, Washington, D.C. Adopted December 11, 1996.

NTSB. Derailment of Union Pacific Railroad Unit Freight Train 6205 West near Kelso, California, January 12, 1997, Report Number: RAR-98-01. Adopted February 6, 1998.

NTSB. Railroad Accident Report, Rear-End Collision of Union Pacific Railroad Company Freight Trains Extra 3119 West and Extra 8044 West near Kelso, California, November 17, 1980, NTSB-RAR-81-7.

NTSB. Railroad Accident Report NTSB/RAR-02/02, Derailment of CSX Transportation Coal Train V986-26 at Bloomington, Maryland, January 30, 2000. Adopted March 5, 2002.

NTSB. Railroad Accident/Incident Summary Report—Derailment of Amtrak Train 87, Silver Meteor, in Palatka, Florida, on December 17, 1991, NTSB/RAR-93/02/SUM. Adopted July 26, 1993.

NTSB. Recommendation, R-93-16, August 17, 1993.

NTSB. Safety Recommendation R-02-8 through 12, March 21, 2002.

"One Goal, Many Game Plans—ECP Brakes—Electro-pneumatic Braking Systems." *Railway Age*, March 1998.

Palmer, Donald E., Henry Christie, and Joe Hettinger. Rail Transportation Division History & Heritage Committee, Oral History of David G. Blaine (1918–2002), Fellow—ASME, ASME International, April 3, 2001.

Peters, A. J., and G. B. Anderson. "A.C. Traction: How Soon?" *Railway Age*, September 1991.

Prout, Henry G. *A Life of George Westinghouse*. New York: Charles Scribner's Sons, 1922.

Prout, Henry G. "Safety in Railroad Travel." *Scribner's Magazine*, September 1889.

Railroad Fire Prevention Field Guide. California Office of the State Fire Marshal Fire Safe Planning, April 26, 1999.

Railroad Friction Products Corporation. Tread Conditioning Brake Shoe Specification (AAR M-997), July 28, 2010.

"The Recent Brake Tests at Burlington." *Railroad Gazette*, October 7, 1887.

Report to Congress on the Large Truck Crash Causation Study, U.S. Department of Transportation, Federal Motor Carrier Safety Administration, MC-R/MC-RRA, Washington, D.C., March 2006.

"Road to Guard Air Brakes." *New York Times*, February 3, 1953.

Rules for Equipment Operation and Handling, NS-1. Norfolk Southern. Effective October 1, 2007.

"Series of Short Blasts Standard Danger Signal." *New York Times*, January 16, 1953.

Shaw, Robert B. *Down Brakes*. London: P. R. MacMillan Limited, 1961.

Technical Specification for the Purchase of Silverliner V Electric Multiple Unit Commuter Rail Cars. Southeastern Pennsylvania Transportation Authority New Vehicle Programs. August 2005.

Traffic Safety Facts. 2006 Data, Large Trucks, DOT HS 810 805, NHTSA's National Center for Statistics and Analysis, March 2008.

"Train Brakes Untried Till Crash in Capital." *New York Times*, January 25, 1953.

Transportation Safety Board of Canada. Derailment on January 8, 2001. Report Number R01W0007. Adopted on February 12, 2003.

Transportation Safety Board of Canada. Runaway/Derailment on December 2, 1997. Report Number R97C0147. Adopted December 10, 1999.

Transportation Safety Board of Canada. Runaway/Derailment on January 18, 1994. Report Number R94V0006. Adopted on February 28, 1995.

Transportation Safety Board of Canada. Runaway/Derailment on June 29, 2006. Report Number R06V0136. Adopted May 20, 2009.

Usselman, Steven W. "Air Brakes for Freight Trains: Technological Innovation in the American Railroad Industry 1869–1900." *Business History Review*, Spring 1984.

Vantuono, William C. "Stop, In the Name of Progress." *Railway Age*, April 2008.

Vantuono, William C. "The Triumph of the 'Tin Horse'—Diesel Locomotives." *Railway Age*, January 1995.

Wald, Matthew L. "After 3 Months on the Shelf, Acelas Begin a Return to Service." *New York Times*, July 12, 2005.

Wald, Matthew L. "Cracks in Brake Prompt Amtrak to Halt Acelas." *New York Times*, April 16, 2005.

Wald, Matthew L. "Official Outlines Roots of Acela Problems." *New York Times*, May 12, 2005.

"The Westinghouse Brake." *Railroad Gazette*, December 2, 1887.

"The Westinghouse Trials." *Railroad Gazette*, November 11, 1887.

Wohleber, Curt. "'St. George' Westinghouse." *Invention & Technology*, Winter 1997.

Ytuarte, Christopher. "Taking a Brake." *Railway Age*, January 2003.

Chapter 11. Broken Rail

"50-Year-Old Freight Cars to Be Banned from Rail Interchange." *Railway Age*, December 1, 1969.

Bayley, Matthew. "The Train Lurched Violently and Turned Over." *Daily Mail*, February 8, 2005.

Bitzan, John D., and Denver D. Tolliver. "North Dakota Strategic Freight Analysis, Item IV. Heavier Loading Rail Cars," Upper Great Plains Transportation Institute, October 2001.

Blaine, D. G., et al. "Train Brake and Track Capacity Requirements for the '80's." *Transactions of the ASME*, November 1981.

Cannon, D. F. *"Joint Research Project 1—Rail Defect Management*, Final Report, Part B, The Synthesis Report" (France's International Union of Railway), June 2003.

Cannon, D. F., et al. "Rail Defects: An Overview." *Fatigue and Fracture of Engineering Materials and Structures*, October 2003.

Cantrell, Darrell D. "Sub-Structure can Add Life to Rail, Ties." *Railway Track and Structures*, March 2003.

Carry, E. F. "Railroad Cars, Their Origin and Development." *Railway Age*, June 23, 1923.

Casavant, K., and D. Tolliver. "2001 Impacts of Heavy Axles Loads on Light Density Lines in the State of Washington," WA-RD 499.1, Washington State Department of Transportation, February 2001.

CN Submission to the Railway Safety Act Review Panel, CN Integrated Safety Plan Technology, May 4, 2007.

Code of Federal Regulations. Title 49: Transportation. *PART 213—TRACK SAFETY STANDARDS*.

Donohue, Pete. "It's Wheely Bad!" *New York Daily News*, March 7, 2005.

English, G. W., and T. W. Moynihan. "Causes of Accidents and Mitigation Strategies, Submitted to Railway Safety Act Review Secretariat," Kingston, Ontario, July 2007.

"Freight Railroads: A Historical Perspective." Association of American Railroads, May 2008.

Garcia, Greg, and David Y. Jeong. "Rail Defect Growth under Heavy Axle Loads." *Railway Track and Structures*, November 2003.

Grain Transportation in the Great Plains Region in a Post-Rationalization Environment: Volume I. Prepared for Bureau of Transportation Statistics, U.S. Department of Transportation, by Upper Great Plains Transportation Institute, December 2005.

Harper, Keith. "Railtrack 'Failed to Repair Track.'" *The Guardian*, October 20, 2000.

"Have Railroad Managers Done a Good Job?" *Railway Age*, May 17, 1947.

Hay, William. *Railroad Engineering*. 2nd ed. New York: John Wiley & Sons, 1982.

Hay, William. *Railroad Engineering*. Vol. 1. New York: John Wiley & Sons, 1953.

"Higher Payloads Proposed by Mechanical Division Officers." *Railway Age*, July 10, 1961.

Jeong, David Y. "Correlations between Rail Defect Growth Data and Engineering Analyses Part I: Laboratory Tests," Volpe National Transportation Systems Center, May 2003.

Jeong, David Y. "Correlations between Rail Defect Growth Data and Engineering Analyses, Part II: Field Tests," Volpe National Transportation Systems Center, January 2003.

Judge, Tom. "Solving Weighty Problems." *Railway Age*, June 2009.

"The Latest Scheme for Pooling Cars." *Railway Age*, March 4, 1922.

Magel, Eric, et al. "Control of Rolling Contact Fatigue of Rails." Proceedings of the AREMA 2004 Annual Conferences, September 19–22, 2004, Nashville, Tennessee.

McGonigal, Robert S. "Rail." *Trains*, May 2006.

Middleton, William D., et al., eds. *Encyclopedia of North American Railroads.* Bloomington: Indiana University Press, 2007.

"The New Railroads: Physically Shrinking, Financially Growing." *Railway Age*, January 1990.

Orringer, O. Control of Rail Integrity by Self-Adaptive Scheduling of Rail Tests, Report No. DOT/FRA/ORD-90/05, June 1990.

Palese, J. W., and T. W. Wright. "Risk Based Ultrasonic Rail Test Scheduling on Burlington Northern and Santa Fe," paper presented at the American Railway Engineering and Maintenance-of-Way Association Annual Technical Conference, September 10–13, 2000, Dallas, Texas.

Parmalee, Julius H. "A Review of Railway Operations in 1930." *Railway Age*, January 3, 1931.

Perry, Keith, and Steven Morris. "Inspection of Site Was Due on Day of Crash." *The Guardian*, October 19, 2000.

Pocket Guide to Transportation. Bureau of Transportation Statistics, U.S. Department of Transportation, January 2009.

"Rail Output Continues Declines in 1922." *Railway Age*, May 5, 1923.

"Rails Broken by Flat Wheels." *Railroad Gazette*, May 10, 1912.

Selig, Ernest T. "Ballast m/w: Putting Basics into Practice." *Railway Track and Structures*, September 2002.

Statistical Abstract of the United States. 1911, thirty-fourth Number. No. 175.—Railroad Freight Cars: Number and Average Capacity of Each Class.

Statistical Abstract of the United States. 1921, forty-fourth Number. No. 243.—Railroad Freight Cars: Number and Average Capacity of Each Class.

Statistical Abstract of the United States 1978. 99th Annual Edition. No. 1122. Railroads—Equipment in Service: 1950 to 1987.

Talbot, A. N. "Report of Committee on Stress in Track." *Railway Age*, March 19, 1938.

Tolliver, Denver. "Eastern Washington Grain-Hauling Short-Line Railroads," Washington State Department of Transportation, February 2003.

Tolliver, Denver, and Alan Dybing. "A Report to the North Dakota Legislative Council," The Upper Great Plains Transportation Institute, April 2007.

Train Derailment at Hatfield: A Final Report by the Independent Investigation Board, Office of Rail Regulation, London, July 2006.

Transportation Safety Board of Canada. Derailment on August 3, 2005. Report Number R05E0059. Adopted on August 20, 2007.

Transportation Safety Board of Canada. Derailment on March 12, 2001. Report Number R01H0005. Adopted December 17, 2002.

Transportation Safety Board of Canada. Derailment on March 21, 1996. Report Number R96T0095. Adopted on November 5, 1997.

"TTCI 15th Annual AAR Research Review." Pueblo, Colorado, March 2–3, 2010.

Turik, Bob. "How Heavy Can Freight Cars Get?" *Trains*, March 2006.

Tuzik, Robert E. "How Well Do You Manage Your Single Biggest Asset?" *Railway Age*, July 1996.

"Two Roads Adopt Heavier Rail." *Railway Age*, August 20, 1921.

Union Pacific. 2009 Annual Report, Omaha, Nebraska.

U.S. Census Bureau. The 2008 Statistical Abstract, The National Data Book, Table 1088, Railroads, Class I—Summary: 1990 to 2007.

Zarembski, Allan M. "Characterization of Broken Rail Risk for Freight and Passenger Railway Operations," paper presented at the AREMA 2005 Annual Conference, Chicago, Illinois.

Zarembski, Allan M., and Joseph W. Palese. "Managing Risk Improves Track Safety." *International Railway Journal*, May 2005.

Chapter 12. Buckled Track

Akhtar, Muhammad N. "Performance Evaluation of Mechanical Rail Joints." *Railway Track and Structures*, July 2008.

American Railway Engineering & Maintenance of Way Association. *Practical Guide to Railway Engineering.* 2nd ed. Landover, Md., 2003.

"Automating Rail Flaw Inspection." *Railway Track and Structures*, Annual 2009.

"A Closer Look at Rail Flaws: All Rail Develops Metallurgical Flaws Sooner or Later." *Railway Track and Structures*. Annual 2008.

Davis, David D. "Rail Joint Installation and Maintenance Recommended Practices." *Railway Track and Structure*, March 2009.

FRA. Research Results, RR08-06, Development of Rail Temperature Prediction Model, June 2008.

FRA. Research Results, RR08-31, Development of Rail Neutral Temperature Monitoring Device, December 2008.

FRA. Track Inspection Time Study. DOT/FRA/ORD-11/15. July 2011.

FRA. Track Safety Standards Compliance Manual, Chapter 5, Track Safety Standards Classes 1 through 5, July 27, 2006.

FRA. Track Safety Standards Compliance Manual, Chapter 5, Track Safety Standards Classes 6 through 9, January 1, 2002.

Gibbons, Timothy J. "Everyone Went Flying and Screaming." *Florida Times Union*, April 18, 2003.

Green, Jim, and Peter Shrubsall. "Management of Neutral Rail Temperature," paper presented at the AREMA 2002 Conference, September 22–25, 2002, Washington, D.C.

Harrison, Harold, Ryan McWilliams, and Andrew Kish. "Handling CWR Thermal Stresses." *Railway Track and Structures*, October 2007.

Hay, William. *Railroad Engineering*. Vol. 1. New York: John Wiley & Sons, 1953.

Igwemezie, Jude, and Anh Tuan Nguyen. "Anatomy of Joint Bar Failures." *Railway Track and Structures*, July 2009.

Jeffrey, Brandon D., and M. L. Peterson. "Assessment of Rail Flaw Inspection Data," MPC Report No. 99-106, Mountain-Plains Consortium, Department of Civil Engineering, Colorado State University, Fort Collins, Colorado, August 1999.

Kish, A., and G. Samavedam. "Dynamic Buckling of Continuous Welded Rail Track: Theory, Tests, and Safety Concepts," Rail: Lateral Track Stability 1991, Transportation Research Board, Transportation Research Record No. 1289.

Kish, A., and G. Samavedam. Dynamic Buckling Test Analyses of a High Degree CWR Track, DOT/FRA/ORD-90/13, February 1991.

Kish, Andrew, and Gopal Samavedam. "Risk Analysis Based CWR Track Buckling Safety Evaluations." Proceedings of the International Conference on Innovations in the Design & Assessment of Railway Track, December 2–3, 1999, Delft University of Technology, The Netherlands.

Kish, A., G. Samavedam, and D. Wormley. "Fundamentals of Track Lateral Shift for High-Speed Rail Applications," paper presented at the European Rail Research Institute's Interactive Conference on "Cost Effectiveness and Safety Aspects of Railway Track," December 8–9, 1998, Paris, France.

NTSB. Derailment of Amtrak Auto Train P052-18 on the CSXT Railroad near Crescent City, Florida, April 18, 2002, NTSB/RAR-03/02. Washington, D.C. Adopted August 5, 2003.

NTSB. Derailment of Amtrak Train No. 6 on the Burlington Northern Railroad, Batavia, Iowa, April 2, 1990, RAR-91-05. Washington, D.C. Adopted December 12, 1991.

NTSB. Derailment of Canadian Pacific Railway Freight Train 292-16 and Subsequent Release of Anhydrous Ammonia near Minot, North Dakota, January 18, 2002, Railroad Accident Report RAR-04/01. Washington, D.C. Adopted March 9, 2004.

NTSB. Railroad Accident Brief, Accident Number DCA-06-FR-006, Arcola, Louisiana, June 26, 2006. Adopted December 21, 2006.

NTSB. Railroad Accident Report, Derailment of Amtrak Train No. 58, *City of New Orleans*, near Flora, Mississippi, April 6, 2004, NTSB/RAR-05/02. Washington, D.C. Adopted July 26, 2005.

Patterson, Steve, and Dana Treen. "Amtrak Train Derails, 6 Dead, Hundreds Hurt." *Florida Times-Union*, April 19, 2002.

Peters, Nigel, et al. "Research and Development of a New Method for Wirelessly Interrogating the Stress Free Temperature of Continuously Welded Rail," paper presented at the AREMA 2008 Conference, September 21–24, 2008, Salt Lake City, Utah.

Rail Adjustment Manual. RTS 3640. Engineering Practices Manual, Civil Engineering. Australian Rail Track Corporation LTD. August 31, 2006.

Railway Accident, Report on the Derailment that occurred on the 5th November, 1967 near Hither Green. Ministry of Transport. Her Majesty's Stationery Office. London, August 8, 1968.

Read, David M., Andrew Kish, and Dwight W. Clark. "Optimized Readjustment Length Requirements for Improved CWR Neutral Temperature Management," paper presented at the AREMA 2007 Annual Conference, September 9–12, 2007, Chicago, Illinois.

Samavedam, G., et al. Parametric Analysis and Safety Concepts of CWR Track Buckling, DOT/FRA/ORD-93/26, December 1993.

Samavedam, G., et al. Safety of High-Speed Ground Transportation Systems, Analyses of Track Shift Under High-Speed Vehicle-Track Interaction, DOT/FRA/ORD-97/02, June 1997.

Samavedam, Gopal, Andrew Sluz, and Andrew Kish. "The Effect of Realignment on Track Lateral Stability," paper presented at the AREMA 1999 Track & Structures Annual Conference, September 12–15, 1999, Chicago, Illinois.

Staff Reporters. "City of New Orleans Train Back on Track." *Times-Picayune*, April 10, 2004.

Transportation Safety Board of Canada. Derailment on August 13, 2002. Report Number R02M0050. Adopted November 13, 2003.

Transportation Safety Board of Canada. Derailment on July 3, 2002. Report Number R02D0069. Adopted February 16, 2004.

Transportation Safety Board of Canada. Derailment on July 4, 2005, Report Number R05H0013. Adopted May 29, 2006.

Transportation Safety Board of Canada. Derailment on July 23, 2002. Report Number R02C0054. Adopted June 2, 2003.

Transportation Safety Board of Canada. Derailment on June 14, 1996. Report Number R96C0135. Adopted February 25, 1998.

Transportation Safety Board of Canada. Derailment on March 29, 2007. Report Number R07D0030. Adopted January 30, 2008.

Varney, James. "Officials Comb Amtrak Wreck." *Times-Picayune*, April 8, 2004.

Wanek, Mischa. "Bettering Ballast." *Railway Track and Structures*, August 2006.

Epilogue. Safety in the Modern Era

About the NTSB: Background, Mission and Mandate, www.ntsb.gov/Abt_NTSB/history.htm.

Garcia, Gregory. "Shattering Conclusions." *Railway Age*, March 2007.

Hansen, Peter A. "6 High-Tech Advances." *Trains*, November 2008.

"High-tech Helps Keep Track of Ballast." *Railway Track and Structures*, March 2004.

Judge, Tom. "A Watchful Eye." *Railway Age*, August 2008.

Kos, Sayre C. "Washout Disasters Are Washed-up." *Trains*, August 2009.

McGonigal, Robert S. "Defect Detectors." *Trains*, May 2006.

TTCI 15th Annual AAR Research Review, Pueblo, Colorado, March 2–3, 2010.

Vanderbilt, Tom. *Why We Drive the Way We Do (and What It Says About Us)*. Knopf, 2008.

"Wheel Profile Measurement." *Railway Track and Structures*, July 2004.

INDEX